ETHICAL ISSUES IN THE 21ST CENTURY

ETHICAL ISSUES IN NEUROSCIENCE RESEARCH

INTEGRATIVE APPROACHES AND PATHS TO PROGRESS

ETHICAL ISSUES IN THE 21ST CENTURY

Additional books in this series can be found on Nova's website
under the Series tab.

Additional e-books in this series can be found on Nova's website
under the e-book tab.

ETHICAL ISSUES IN THE 21ST CENTURY

ETHICAL ISSUES IN NEUROSCIENCE RESEARCH

INTEGRATIVE APPROACHES AND PATHS TO PROGRESS

MARILYN PHILLIPS
EDITOR

nova publishers
New York

Library of Congress Cataloging-in-Publication Data

ISBN: 978-1-63482-989-2

Published by Nova Science Publishers, Inc. † New York

CONTENTS

PREFACE

This book is a set of recommendations from the Bioethics Commission in response to a request from President Obama to review the ethical issues associated with the conduct and implications of neuroscience research; and President Obama's request related to the Brain Research through Advancing Innovative Neurotechnologies (BRAIN) Initiative. Specifically the President asked the Bioethics Commission to "identify proactively a set of core ethical standards – both to guide neuroscience research and to address some of the ethical dilemmas that may be raised by the application of neuroscience research findings." This book focuses on the integration of ethics into neuroscience research across the life of a research endeavor; and on the analysis on three particularly controversial topics that illustrate the ethical tensions and societal implications of advancing neuroscience and technology: cognitive enhancement, consent capacity, and neuroscience and the legal system. The book seeks to clarify the scientific landscape, identify common ground, and recommend ethical paths forward.

In: Ethical Issues in Neuroscience Research
Editor: Marilyn Phillips ISBN: 978-1-63482-989-2

Chapter 1

GRAY MATTERS: INTEGRATIVE APPROACHES FOR NEUROSCIENCE, ETHICS, AND SOCIETY*

Presidential Commission for the Study of Bioethical Issues

President Barack Obama
The White House
1600 Pennsylvania Avenue, NW
Washington, DC 20500

Dear Mr. President:

On behalf of the Presidential Commission for the Study of Bioethical Issues, we present to you *Gray Matters: Integrative Approaches for Neuroscience, Ethics, and Society*, the first part of the Bioethics Commission's response to your request of July 1, 2013. The Bioethics Commission resolved to address your charge in multiple reports to be timely and responsive in considering the rapidly emerging and evolving field of neuroscience. In this first report, the Bioethics Commission considered the need for, and various approaches to achieve, integration of ethics early and throughout neuroscience research. Explicit integration of ethics and neuroscience research will help researchers, policymakers, and the public recognize and address the ethical and societal implications of neuroscience research and its applications. The Bioethics Commission will consider the societal implications in greater detail in a later report.

Building on its past work on neuroscience and related ethical issues, which included hearing from experts at two public meetings, the Bioethics Commission held three public meetings on this topic and heard from speakers from a variety of disciplines and perspectives, including neuroscientists, philosophers, educators, ethicists, and representatives from federal agencies as well as private sector partners involved in the Brain Research through Advancing Innovative Neurotechnologies (BRAIN) Initiative. In addition, the Bioethics Commission solicited public comment and received many thoughtful responses.

* This is an edited, reformatted and augmented version of Volume 1 of a two volume report, issued May 2014.

Contemporary neuroscience investigates the human brain and its relation to the mind, raises fundamental questions about human experiences, and has produced and could yield improved preventive, diagnostic, and therapeutic interventions for brain-related diseases that affect tens of millions of Americans. Ethical issues arise in neuroscience research, similar to other scientific fields, but are sometimes expressed in sharper relief. Ethics integration should equip scientists to recognize and address ethical issues as they arise, and ethicists to understand the science and technology with which they engage.

The Bioethics Commission concludes that ethics and neuroscience research should be integrated throughout the research endeavor and offers four recommendations to facilitate the successful integration of ethics and neuroscience research. Institutions and individuals should take steps to make explicit their plans for integrating ethics across the research endeavor, and institutions should provide sufficient resources to support integration efforts. Institutions and researchers should evaluate existing and innovative approaches to integrating ethics and neuroscience, including integration through education at all levels. In addition, professionals with experience in ethics should be included in BRAIN Initiative-related scientific advisory boards and funding review committees, particularly for the major public and private sector partners.

The Bioethics Commission is honored by the trust you have placed in us and we are grateful for the opportunity to serve you and the nation in this way.

Sincerely,
Amy Gutmann, Ph.D.
Chair

James W. Wagner, Ph.D
Vice Chair

THE WHITE HOUSE

WASHINGTON

July 1, 2013

The Honorable Amy Gutmann, Ph.D.
Commission Chair
Presidential Commission for the Study of Bioethical Issues
Washington, D.C. 20005

Dear Dr. Gutmann:

As I noted in my announcement of the Brain Research through Advancing Innovative Neurotechnologies (BRAIN) Initiative on April 2, 2013, developments in neuroscience hold great potential to help individuals and society. New technologies to better visualize the brain and understand how it works promise to speed the discovery of new ways to treat and prevent brain disorders, including those caused by disease and traumatic injury, and to shed light on the neural components of memory and learning, among other benefits.

Advances in neuroscience can also raise ethical and legal issues that require reflection and analysis. In keeping with my Administration's strong commitment to rigorous research ethics in all fields, I want to ensure that researchers maintain the highest ethical standards as the field of neuroscience continues to progress. As part of this commitment, we must ensure that neuroscientific investigational methods, technologies, and protocols are consistent with sound ethical principles and practices.

Equally important, we should consider the potential implications of the discoveries that we expect will flow from studies of the brain, and some of the questions that may be raised by those findings and their applications—questions, for example, relating to privacy, personal agency, and moral responsibility for one's actions; questions about stigmatization and discrimination based on neurological measures of intelligence or other traits; and questions about the appropriate use of neuroscience in the criminal-justice system, among others. It will also be important to consider these types of questions as they relate to different life stages, from infancy through old age.

I request that the Presidential Commission for the Study of Bioethical Issues engage with the scientific community and other stakeholders, including the general public, to identify proactively a set of core ethical standards—both to guide neuroscience research and to address some of the ethical dilemmas that may be raised by the application of neuroscience research findings.

In the course of your deliberations, I encourage you to reach out to a wide range of constituencies, including scientists, ethicists, legal scholars, and members of the public, to ensure that your findings and the neuroscience enterprise faithfully reflect and strengthen our values as a Nation.

Sincerely,

Report and Recommendations

About This Report

This first report by the Presidential Commission for the Study of Bioethical Issues (Bioethics Commission), as part of the Brain Research through Advancing Innovative Neurotechnologies (BRAIN) Initiative, emphasizes the importance of integrating ethics into neuroscience research from the earliest planning phases. Given the promise and potential of neuroscience, we are compelled to consider carefully scientific advances that could alter our conception of the very private and autonomous nature of self. Our understanding of the mind, our private thoughts, and our volition necessitates careful reflection about the scientific, societal, and ethical aspects of neuroscience endeavors. Integrating ethics explicitly and systematically into the relatively new field of contemporary neuroscience allows us to incorporate ethical insights into the scientific process and to consider societal implications of neuroscience research from the start. Early ethics integration can prevent the need for corrective interventions resulting from ethical mishaps that erode public trust in science.

This report provides practical, conceptual, and methodological tools that can be applied directly in neuroscience research by funders, scientists, and other stakeholders. It calls for adequate resources to be allotted for successful integration of science and ethics. It also provides analysis and recommendations to guide institutions in developing necessary infrastructure for early integration of ethics into neuroscience research. In a second report, the Bioethics Commission will consider the ethical and societal implications of neuroscience research and its applications more broadly—ethical implications that a strongly integrated research and ethics infrastructure will be well equipped to address, and that myriad stakeholders, including scientists, ethicists, educators, public and private funders, advocacy organizations, and the public must be prepared to handle.

* * *

In July 2013, President Obama charged the Bioethics Commission to consider ethical and societal issues associated with advances in the field of neuroscience.[1] The Bioethics Commission is addressing the President's charge in multiple reports to be timely and responsive in the face of a rapidly emerging and evolving field. This first report reflects the Bioethics Commission's commitment to early integration of ethics and scientific research through rapid dissemination of the insights it has gained and the conclusions it has reached.[2]

Anatomist and Nobel laureate Santiago Ramón y Cajal wrote: "[t]he brain is a world consisting of a number of unexplored continents and great stretches of unknown territory."[3] Cajal's perspective, pronounced a century ago, remains as true today as then. The BRAIN Initiative, with its Presidential endorsement and initial investment of federal funding, represents, both practically and symbolically, an exciting investment in basic neuroscience research. Focusing on the development and application of new technologies to visualize and understand complex neural circuits, the BRAIN Initiative has the potential to make major contributions to knowledge about how the brain works.

This knowledge, in turn, could inform the development of better methods for preventing and treating neurological conditions. President Obama asked the Bioethics Commission to examine the ethical considerations of not only BRAIN Initiative-related research, but neuroscience research more generally. With this broad focus, the recommendations in this report are relevant for the integration of ethics and science in neuroscience and other scientific fields.

Ethics and Neuroscience Research

The human brain has long been a focal point of scientific and public inquiry and concern. Contemporary neuroscience—which includes molecular neuroscience, cognitive neuroscience, and clinical neuroscience, among many other subfields—has produced and could yield improved preventive, diagnostic, and therapeutic interventions for conditions that affect tens of millions of Americans and as many as a billion people worldwide.[4] Advances in neuroscience also can raise fundamental questions about central features of human experience such as conscious awareness and personal identity.

Burden of Neurological Disorders in the United States and Worldwide

In the United States and worldwide, the burden of neurological disorders is high and projected to "increase considerably in years to come." Neurological disorders are estimated to affect as many as one billion people globally, and millions of people in the United States. These disorders affect all age groups across geographical regions. There are many neurological illnesses with substantial health and public health impacts including dementia, epilepsy, multiple sclerosis, Parkinson's disease, schizophrenia, and traumatic brain injury, for which neuroscience research has the potential to contribute to improved interventions.

Sources: World Health Organization (WHO). (2006). *Neurological Disorders: Public Health Challenges*, Chapter 4: Conclusions and Recommendations, p. 177. Retrieved April 4, 2014 from http://www.who.int/mental_health/neurology/ chapter_4_neuro_disorders_public_h_ challenges.pdf; Collins, P.Y., et al. (2011). Grand challenges in global mental health. *Nature*, 475(7354), 27-30; Hirtz, D., et al. (2007). How common are the "common" neurologic disorders? *Neurology*, 68(5), 326-337; Lozano, R., et al. (2012). Global and regional mortality from 235 causes of death for 20 age groups in 1990 and 2010: A systematic analysis for the Global Burden of Disease Study 2010. *Lancet*, 380(9859), 2095-2128; Murray, C.J.L., et al. (2013). Disability-adjusted life years (DALYs) for 291 diseases and injuries in 21 regions, 1990-2010: A systematic analysis for the Global Burden of Disease Study 2010. *Lancet*, 380(9859), 2197-2223.

Many of the ethical issues raised by neuroscience research are not unique.[5] The ethical issues surrounding data privacy, informed consent, and minimization of risk, for example, are common across scientific fields. Some issues, however, such as those regarding privacy of our thoughts, threats to personal volition, or erosion of self-determination are expressed in sharper relief in neuroscience.

Advances in neuroscience highlight the relationship between theory and practice, between what we think and feel, and what we do. The complexity of the relationships among human thought, emotion, and action captures the importance of neuroscience to individuals and society. Conscious experience—communicated in many individual and cultural ways—is undoubtedly a central defining feature of human existence, and is intricately connected to human emotion and action. Advances in neuroscience excite public curiosity and challenge our self-understanding more than scientific inquiries into other bodily organ systems that lack the cultural significance of the brain as a prime locus of personal identity.[6] Grappling with complex issues that have theoretical, practical, and social dimensions suggests a need for deliberate, thoughtful, and inclusive engagement to develop responsive and nuanced guidance.

The purpose of this first report is to underscore the importance of integrating ethics and neuroscience early and throughout the research endeavor. Such integration offers a means by which researchers can recognize and respond to ethical issues that arise throughout the research process. Timely recognition of ethical issues allows one to remain responsive to the progress of research. As a starting point, this report highlights some of these ethical and societal issues that can arise through the introduction of four examples to demonstrate the need for integrating ethics throughout neuroscience research: (1) neuroimaging and brain privacy; (2) dementia, personality, and changed preferences; (3) cognitive enhancement and justice; and (4) deep brain stimulation research and the ethically difficult history of psychosurgery. While these examples bring pressing ethical issues to the fore, they will not be extensively analyzed or resolved in this report. The Bioethics Commission's second report on neuroscience and ethics will consider in greater detail the ethical and societal implications of neuroscience and the products of neuroscience research, including, for example, how neuroscience research fi ndings are communicated.[7]

Neuroimaging and Brain Privacy

A variety of neuroimaging technologies, including computed tomography (CT), positron emission tomography (PET), electroencephalography (EEG), and functional magnetic resonance imaging (fMRI) enable us to see the brain or its activity in different ways. New insights into the brain can raise questions about privacy, particularly concerning individuals' ability to decide whether and how personal, sensitive, or intimate information is acquired and used.[8] The potential use of neuroimaging techniques for crime prevention, lie detection, or to make inferences about criminal intent raises broader concerns about the societal implications of the underlying research.[9]

The possibility of using neuroscience for such purposes is already being explored. Several laboratory studies, for example, have used fMRI to investigate questions about deception.[10] In one study, investigators asked research participants to undertake a "mock crime" by stealing either a ring or a watch. Participants were then instructed to deny stealing either of the objects, and underwent fMRI while responding to a series of questions, some of which were intended to prompt the instructed false denial. The investigators then used the fMRI data of the truthful and untruthful responses to build models to detect the deception. In the end, one model correctly predicted the veracity of responses in 90 percent of participants.[11] The investigators underscored one key limitation: "[t]his study...did not evaluate real-world scenarios or lies with severe societal, emotional, or monetary damages."[12]

This example highlights concerns about the potential—whether real or imagined—of technologies such as fMRI to observe private thoughts. As we anticipate personal and societal implications of using such technologies, ethical considerations must be further deliberated.

Dementia, Personality, and Changed Preferences

Dementia is a large and rapidly growing public health concern affecting approximately 2.4 to 5.5 million Americans and 35.6 million people worldwide.[13] It encompasses a group of cognitive disorders characterized by memory impairment and difficulty with language, motor activity, object recognition, and executive function.[14] The prevalence of dementia increases with age, affecting 5 percent of persons 71-79 years of age, 24 percent of persons 80-89 years of age, and 37 percent of persons over 90 years of age.[15]

Dementia raises ethical and societal issues with which neuroscience must engage.[16] With progress in the ability to predict future disease, growing numbers of individuals at risk might use this information to plan for their futures.[17] Future planning for disease conditions in which a person will lose the capacity to make decisions for themselves is not uncommon in health care, but it raises difficult philosophical and ethical issues.[18] Prominent among these are questions about how dementia affects notions of the self. Because the neurological changes caused by dementia can be profound, patients' wishes and preferences might change considerably as their dementia worsens and they progress from non-impaired to severely impaired states. Whereas the references expressed by an individual in a non-impaired state (sometimes formally expressed in an advance directive) are those of an autonomous person, the same individual with dementia might express preferences that differ from previous preferences.

Consider, for example, a man who, prior to having dementia, expressed a strong and clear preference against invasive treatments if his cancer were to return. The cancer recurs many years later after his dementia has progressed, at which time he insists on a major surgery that might extend his life, but is unlikely to markedly improve its quality. The ethical question of whether to move forward with the surgery focuses on whether his pre-dementia or post-dementia preferences should take priority and which decision best reflects his overall wishes and best interests.

Philosophers and ethicists have taken several positions in relation to cases like this one. Some have argued that respecting an individual's autonomy requires following their earlier autonomous preferences.[19] Others have argued that substantial weight should be given to the individual's current preferences, which might be markedly different from an earlier point in time.[20]

Health care professionals, caregivers, and loved ones must reconcile the ethical difficulties they encounter when pre-dementia autonomy, post-dementia autonomy, decision making capacity, and beneficence (that is, acting to ensure the wellbeing of others) appear to conflict.[21] Neuroscience, in its focus on the brain, makes engagement with these issues inevitable, as clinicians, researchers, family members, and others interact with and seek to improve the lives of those affected by neurological disorders; its discoveries also could change how we view the self of those with dementia, and our views of selfhood in general.

Cognitive Enhancement and Justice

Neuroscience advances can raise questions about justice. Humans have long had hopes and concerns about (and used) pharmacological and technological interventions meant to improve mental and physical capacities beyond normal functioning. The term "cognitive enhancement" covers a broad spectrum of methods—from nutritional supplements to brain stimulation—that improve or are thought to improve some aspect of cognitive function.[22] Coffee consumption and yoga, for example, are used by some for their cognitive effects. Certain scholars contend that some forms of cognitive enhancement are socially acceptable or possibly even desirable.[23] This example focuses primarily on prescription stimulant drugs that might be used for enhancement purposes.

Many pharmaceutical interventions associated with cognitive enhancement were developed originally for traditional clinical uses, such as the treatment of attention deficit hyperactivity disorder.[24] A considerable literature in philosophy and bioethics has explored the ethical implications of distinguishing the use of the same intervention as a treatment for a diagnosed medical condition in one case and as an enhancement of non-diseased capacity in another. Some have argued that the distinction matters ethically because while we are obliged to provide people with medical care (treatment) as a matter of justice, we are not obliged to provide people with enhancements.[25] However, beyond the distinction between treatment and enhancement, cognitive enhancement raises pressing ethical issues concerning equitable access to enhancements and their benefits, appropriate management of risks, and obligations and freedoms to enhance or not.[26]

There are anecdotal reports, media accounts, and scientific studies suggesting the use of pharmacological cognitive enhancement, particularly in educational settings such as college campuses.[27] In a web-based survey of medical students, 18 percent reported ever using stimulants, over half of which reported their first use in college.[28] Of those who reported ever using stimulants, 95 percent reported that they perceived that stimulant use could result in "some improvement in academic standing."[29] Among all survey respondents, 11 percent reported stimulant use during medical school. In a 2013 survey, about 10 percent of male and about 8 percent of female undergraduates reported using a prescription stimulant not prescribed to them in the past 12 months.[30] Additionally, in a 2005 survey, college student respondents were asked "how often, if ever, have you used any of the drugs listed below (Ritalin, Dexedrine, or Adderall)? Do not include anything you used under a doctor's orders."[31] Reported use in the past year was higher among students who were male, white, members of fraternities and sororities, had lower grade point averages, and attended more selective colleges.[32] A more recent survey indicates that "non-medical" use of Adderall is increasing, and demographic disparities remain.[33] These data could raise concerns about justice and equity, insofar as using stimulants in this way might be viewed as conferring or reinforcing advantage and exacerbating existing educational, economic, and other disparities.[34]

Deep Brain Stimulation Research and the Ethically Difficult History of Psychosurgery

The final example demonstrates how one type of ethics integration can help to further scientific advancement by addressing previous ethical controversies concerning analogous technologies. Deep brain stimulation (DBS) involves a surgical procedure to implant electrodes into specific regions of the brain for calibrated stimulation of those regions as a therapeutic treatment for symptoms of certain illnesses. The U.S. Food and Drug Administration first approved DBS in 2002 as a treatment for Parkinson's disease.[35] The relative success of DBS to manage many of the motor symptoms of Parkinson's disease, and preliminary research indicating it might be effective for several psychiatric illnesses, suggest that DBS might have a promising future for application in other disorders.[36] Careful research is required to determine whether DBS constitutes a safe and effective treatment for severe psychiatric illnesses, such as treatment-resistant depression and obsessive-compulsive disorder.

Much as the field of genetics had to contend with the history of eugenics, contemporary advances in psychosurgery, such as DBS, must contend with the widespread cultural salience of a controversial past.[37] Psychosurgery generally, and frontal lobotomy in particular, has a controversial history as early "brain science" and treatment for mental illness.[38] Implemented at a time when there were few acceptable or effective treatments for mental illness, lobotomy, at its worst, was characterized by dubious consent, crude technique, lack of scientific evidence, and major side effects.[39] With more than 40,000 procedures performed in the United States during the 1940s and 1950s, lobotomy was held in such high esteem that pioneering work in its development led to the awarding of a Nobel Prize in 1949.[40] At that time, the risks of the procedure were well known but infrequently described in the scientific literature, while its benefits were lauded in popular media. Despite eventual public and professional outcry about ethical concerns, the advent of pharmacological alternatives was likely the greatest driving force in the decline in lobotomies.[41]

A detailed history of the practice of lobotomy describes the promise it held for psychiatric professionals in terms that could now apply equally to DBS as "a ray of hope in an otherwise bleak situation for mentally ill patients."[42] Reflection on this history points out the importance of addressing ethical issues then and now, as "last-ditch medical interventions will probably always be with us. We must therefore continue to scrutinize them, not only in retrospect but as they are being conceptualized, publicized, and carried out."[43]

Because of psychosurgery's culturally and socially sensitive history, scientists and ethicists have done extensive work to safeguard the ethical conduct of current and future DBS research.[44] Proponents of DBS to treat psychiatric illness are highly invested in conducting scientifically and ethically sound research. Bioethicists, patient advocates, policymakers, psychiatrists, neurologists, and other experts met in 2007 to establish consensus regarding the standards and protections that should be in place to facilitate the ethical practice of DBS research.[45] They established a set of ethical safeguards, including the careful design of research protocols; particular attention to the vulnerabilities of participants with debilitating psychiatric illnesses; focus on adequate and fully informed consent; and careful management of expectations regarding the potential for clinical use, thereby mitigating therapeutic misconception (the mistaken conflation of the goals of research and clinical care).[46]

Integrating ethical considerations into future research protocols reduces the likelihood of ethical pitfalls and provides assurance to the public that scientific progress will not be impeded by failure to consider ethical and societal implications. Thoughtful consideration of ethical issues increases the acceptability of DBS research, and thus enhances the likelihood that DBS research will continue in a way that might result in new therapies for individuals suffering from certain psychiatric illnesses.[47]

* * *

Neuroscience is a rapidly growing, multidisciplinary field including physical and biological sciences, behavioral and social sciences, clinical research, engineering, computer science, and mathematics and statistics, each with its own ethical tradition. Integrating these various ethical traditions, elucidating a platform for ethical understanding and discussion, and allowing different ethical perspectives to be emphasized at different stages of a research project can strengthen both implementation of the study and translation of the fi ndings to the broader public.

> "By virtue of having a brain, you have a memory. By virtue of having a memory, you have a personal narrative. And these are the things that are devastated by Alzheimer's and we are trying to understand what goes wrong. In order to do that, we need to go to the physiology, and we need to be able to develop these new neural techniques. And that makes it a very much more challenging enterprise than helping the heart or the liver, all... important for saving lives. But somehow when you start tinkering with the brain, people get a little bit more concerned because it's really tinkering with who you are. And that's I think something that we're all grappling with."

Sejnowski, T., Francis Crick Chair, Professor and Laboratory Head, Computational Neurobiology, Salk Institute for Biological Studies; Investigator, Howard Hughes Medical Institute; and Distinguished Professor, Section of Neurobiology/Neurosciences, University of California, San Diego. (2013). Presentation to the Presidential Commission for the Study of Bioethical Issues, December 18. Retrieved April 4, 2014 from http://bioethics.gov/node/3239.

The many branches of neuroscience address a wide range of topics including but not limited to the evolution of the nervous system (evolutionary neuroscience), the neural mechanisms of language (neurolinguistics), and the neural mechanisms of emotion (affective neuroscience). It is unsurprising, then, that neuroscience encompasses a diversity of scientific and ethical issues.[48] Many neuroscientists explicitly wrestle with the ethical and social implications of their research and how best to address them. Others might grapple with ethical issues, even if they do not label them as such. And some scientists might not see themselves as encountering ethical issues as a part of their work at all.

It is critical that those involved in neuroscience, including researchers, research institutions and universities, public and private funders, and others, have the tools to identify and address the ethical issues they encounter. Without such tools, ethics integration is likely to be viewed by some as irrelevant or impractical, rather than as intrinsic to good science. As a first step to meaningful ethics integration, those involved in neuroscience research must identify the many areas in which ethical questions might arise in relation to their specific roles and responsibilities.

Individual scientists, for example, face ethical questions when they choose research topics, seek funding, design and conduct research, and disseminate results. There are a number of ways they might prepare to address or arrive at solutions to ethical questions, such as through professional education or discussions with experienced colleagues. Universities encounter ethical issues because they fund research; provide laboratories, equipment, and staff; and shape careers through incentive structures such as tenure. Universities can foster a community focused on responsible research through their choices about curricula and faculty, criteria for promotion, and by implementing ethics education for faculty, staff, and students. Funders require approaches that help them address ethical questions in relation to their position to influence research priorities more broadly. These approaches might include supporting research on the ethical and societal implications of a particular line of inquiry or involving those with experience in ethics on panels reviewing grant applications.

Individuals and institutions should assess their roles and influence in neuroscience research and ethical issues they are likely to encounter in order to determine which approaches to ethics integration are best suited to their context.

Approaches to Ethics Integration in Neuroscience

Ethics integration is a process by which scientists and ethicists engage with each other, and often other stakeholders, such as communities, to understand the social and ethical dimensions of their work, including the relationship between science and the societal context in which it operates. Executed well, ethics integration is an iterative and reflective process that enhances both scientific and ethical rigor. Without ethics integration, neuroscience and neuroscientists might overlook fundamental ethical and social dimensions of the complex phenomena they seek to understand. And if ethicists are not conversant in the *science* of neuroscience, they will be unable to make a meaningful contribution to the *ethics* of neuroscience. As an early bioethicist wrote, "integration can only come about by a continuing multidisciplinary discussion between people who are experts in their chosen discipline but who, in addition, have some degree of competence in fields other than their own."[49] Without the other, ethics or neuroscience alone can offer only incomplete guidance and solutions.

Ethics is more than regulatory compliance or risk mitigation, and integration involves a deep collaboration between ethics and science such that the contributions from all disciplines are strengthened. As one presenter to the Bioethics Commission said, "bioethicists and scientists need one another," requiring both to be sufficiently knowledgeable about the others' fields.[50] This mutual understanding fosters credibility and appreciation of the technical aspects and practical exigencies of each other's work, and, for ethicists, helps to avoid overestimating or missing ethical problems.[51] Knowledge of the technical aspects of another discipline, however, does not mean that one abandons the lens of one's original field. All involved must consciously adopt and maintain a critical perspective to mitigate the danger of losing the valuable aspects of one's independent perspective or disciplinary identity.[52]

Ethics As a Lens

In considering ethics integration, and how ethics can facilitate the work of science, the Bioethics Commission discussed the notion of ethics as a lens. Vice Chair Wagner invoked the analogy of the importance of telescope optics in the central work of astronomy.

"[A]stronomers are not physicists or optics designers, but they understand that [the] optics of a telescope is an important companion discipline that both restricts their ability to look at what they're looking at and about which they also hope it will improve."

Similarly, ethics is a necessary companion to all good science that both scientists and ethicists should aim to improve.

Source: Wagner, J.W., Vice Chair, Presidential Commission for the Study of Bioethical Issues; and President, Emory University. (2014). Research and Integration Update and Member Discussion, February 10. Retrieved April 7, 2014 from http://bioethics.gov/ node/3372.

Relevant ethical considerations, and the type and intensity of integration required, will vary depending on the type of research an investigator is conducting. Developing new imaging technologies, analyzing biobank data, or conducting Phase III trials to evaluate the efficacy of a new therapy for a disease will demand a level of integration appropriate to the ethical demands of the context. Some researchers, upon reflection, might assert that they rarely confront difficult ethical issues in their daily work, while the nature of others' investigations might require more explicit, frequent, and intensive engagement with complex ethical questions. Yet relevant ethical issues take many forms based in professional ethics, research ethics, and the societal implications of the knowledge or technologies that research produces (including clinical implications). For example, laboratory researchers studying neurodegeneration in yeast models might routinely encounter questions of professional ethics, such as those related to analytic integrity, authorship, or mentorship of postdoctoral fellows.[53] Those conducting research with human participants have another critical set of ethical concerns with which to grapple. And all researchers must consider the potential societal implications of their research, such as how the results of a well-intended study might yield unintended applications.

Foundational Domains of Ethical Conduct

The foundations for ethical neuroscience research are already in place in the tenets of professional ethics and research ethics, and in early consideration of the societal implications of neuroscience advances. Integration infuses existing codes and models for ethical conduct into the research process from the outset, ensuring that those engaged with neuroscience research are aware of and draw upon these and other applicable ethics frameworks.

Professional ethics in science is derived from expectations of responsible conduct of research, codes of conduct, and the character traits that are the hallmark of good scientists. Professional ethics applies throughout the research process, from the reflective articulation of a research question to the honest and responsible communication of scientific findings.

Research ethics consists of ethical and regulatory guidelines that govern research, including those concerning research involving humans or animals. Responsible neuroscience includes recognition, interpretation, and application of existing ethical principles and regulations; assurance of compliance with regulations; and consideration of other ethical safeguards for human participants and nonhuman animals in research.

Finally, it is likely that a great deal of neuroscience research will have *societal implications*, including clinical implications, even if findings do not translate immediately into new interventions or technologies. Ethical research requires that scientists consider potential societal implications of their work from the outset.

A fundamental goal of scientific inquiry is to seek improved explanations of the world.[54] The potential of science to improve human welfare grounds a societal obligation to undertake and support scientific research generally.[55] Scientists pursue knowledge both for its own sake and for the practical problems it can help to solve. Their professional role grounds a basic duty to pursue science for the public good and to consider carefully the possible consequences of their work.[56] The duty to pursue science in the public interest derives from several ethical obligations, as well as the fact that much scientific research is directly or indirectly publicly funded.

"I think by asking the fundamental question[s]—[W]hy are you doing this? Why do you think your lab director wants to get this grant in particular? Why does your institution promote it in this way? Why did the funding agency write the solicitation in the following way?—by asking these questions you can really impact what's already there, and in the process the scientists can take ownership rather than the moral expert insisting on what the logical case is. ...[I]t's a subtle move, but it allows for a co-responsible approach....

I would suggest that the word 'curiosity'... is potentially an engine for both ethical care and scientific creativity."

Fisher, E., Associate Director for Integration, Center for Nanotechnology in Society; Assistant Professor, School of Politics and Global Studies and the Consortium for Science, Policy and Outcomes, Arizona State University. (2014). How the Inclusion of an Ethicist on a Research Team Might Affect Change in Scientific Research. Presentation to the Presidential Commission for the Study of Bioethical Issues, February 11. Retrieved April 7, 2014 from http://bioethics.gov/node/3379.

Including ethics in science is not a new process. Value decisions implicit in science encompass practical and moral considerations, as experienced researchers appreciate. These include how to allocate limited funds to a number of possible areas of study, determine which research designs will be most beneficial and least harmful to individuals, and identify which experiments are most likely to bring about the most useful results for society at the least cost. Practical and moral considerations come into play as well in the implementation and conduct of research, from the day-to-day work of completing experiments to data analysis.

The purpose of explicitly integrating ethics in science is to engage in ethical analysis and reflection and bring ethical decisions and assumptions inherent to the practice of science to the forefront to assess their merits, develop new standards or modify old ones, and reform practices where needed. Democratic deliberation, an approach to collaborative decision making that embraces respectful debate of opposing views and active participation by citizens, is critical to ethics integration because in an open society—and considering the societal investment in modern science— societal values need to constitute a genuine part of these decisions.[57]

Science driven by "ethics" is not always enough, however, even with public input. For example, some leading scientists, captivated by early 20th century eugenic thought, were driven by an ethical framework of perceived societal benefit.[58] They wrote at length in well-recognized journals and popular publications about the importance of making sure that science was driven by a certain ethos, one that turned out to be catastrophic both for science and for the rest of the world. Modern society can and must learn from these mistakes; this is the nature of the moral life in science and in all human experience. Ethics integration acknowledges that measures can be taken to ensure the likelihood that modern neuroscience will benefit from an awareness of the past without being paralyzed by it. With scientific and public interest in neuroscience growing, the time is especially right for a more explicit conversation about practices that facilitate ethical reflection in the research process.

The following are selected approaches to ethics integration.

Ethics Integration through Education at All Levels

Integration of ethical issues in neuroscience education is clearly demanded by the vibrant nature of the field.[59] Scholars have suggested that in order to develop an understanding and knowledge base in ethics, future research professionals should be exposed to ethics early in their education and careers.[60] This early exposure facilitates a deeper understanding of the implications of science, before students are immersed in their own work and overwhelmed by early career pressures. Integration of ethics and science through education can equip scientists to recognize and assess ethical issues in their work, and equip bioethicists to understand scientific practice.[61]

Early ethics education provides a strong foundation for students who might later pursue science as a profession, and professional and continuing ethics education builds on that foundation for more experienced scientists. Science programs, including but not limited to neuroscience, can challenge students to confront big questions, which mentors and educators should revisit periodically and seriously, such as: "What is science for? What are the values I bring to my scientific work? Why did I become a scientist and why am I one now?"[62] One desirable goal for ethics education is "transformational learning," which goes beyond cultivating cognitive learning or critical thinking to inculcate "habits of mind, attitudes, and dispositions."[63] If successful, it will become natural for scientists to have ongoing engagement with the societal dimensions of their work.[64]

"[B]rain science is a fine place to begin to figure out how to integrate bioethics education into the preparation of scientists. Ethics education in this area will be particularly challenging, and therefore exciting to work on, because brain science requires collaboration across so many disciplines and because it will raise profound questions across all three domains: responsible conduct of research, research ethics, and the societal impact of the knowledge and technologies that emerge. Since existing bioethics education programs have focused much less on this third area of societal impact, and since brain science engages so many questions in that domain, I recommend that there be considerable attention to the ethical and social impact questions, not just to research ethics and RCR [responsible conduct of research]. It is also my hope that we will not just train or educate, but that we will commit to designing for learning, and specifically for a kind of learning that is transformational, so that we are preparing not just scientists, but citizen-scientists who are professional in the fullest meaning of the word, aware of the power science holds in society, and capable of secular moral reasoning in our highly pluralistic society."

Solomon, M.Z., President, The Hastings Center; and Clinical Professor of Anaesthesia, Harvard Medical School. (2013). Transformational Bioethics Learning in Brain Science. Presentation to the Presidential Commission for the Study of Bioethical Issues, December 18. Retrieved April 4, 2014 from http://bioethics.gov/sites/default/files/ Solomon%20Remarks%20to%20 Pres%20Commission.pdf.

Integration through education is not limited to exposing science students to ethics. Ethics programs also can require exposure to scientific methods. Students could be required to complete, for example, a laboratory science component or conduct independent scientific research to enhance their scientific literacy and understanding.

Early integration of ethics into curricula can begin before the undergraduate level, during high school or sooner. For example, in a pilot program that tested the effectiveness of a new high school bioethics curriculum, researchers found that the addition of bioethics awakened a curiosity in students about the science itself; the addition of bioethics did not fetter interest in science, rather, it encouraged it.[65] Efforts such as Brain Awareness Week, a worldwide "celebration of the brain for people of all ages," which includes events intended to educate about neuroscience as well as advocate for science funding, also offer opportunities for education at many levels including the public at large.[66]

Educational efforts can continue throughout higher education. Some colleges and universities in the United States and abroad have existing or are developing programs, courses, or teaching resources focused on ethical issues related to neuroscience. For example, Duke University's graduate program in Bioethics and Science Policy offers a neuroscience concentration and focuses on complex problems at the intersection of science, technology, and ethics.[67] The University of Pennsylvania Center for Neuroscience and Society provides a repository of high school, undergraduate, and graduate level neuroethics teaching materials.[68] And the University of Manchester Interdisciplinary Network on Teaching of Ethics for Neuroscientists developed an online neuroethics education module.[69]

Ethics integration through education can continue still further, through graduate and professional education and throughout a scientific career, for example, via continuing education programs for experienced researchers.

> "I think that science education has to be fundamentally restructured from the get-go, and that that is what is ultimately going to address these issues, and I think that scientists think the way they think because their education leads them to think that way. They're siloed because that's how they're trained. It should be no surprise that they exist within a particular world when that is how science education is organized, and I think that that is something that has to be dealt with."
>
> Sankar, P., Associate Professor, Department of Medical Ethics and Health Policy; and Senior Fellow, Leonard Dan's Institute of Health Economics, University of Pennsylvania. (2014). ELSI Origins and Early History. Presentation to the Presidential Commission for the Study of Bioethical Issues, February 11, Retrieved April 4, 2014 from http://bioethics.gov/node/3379.

Ethics Integration through Institutional Infrastructure

Public and private institutions can develop structures and processes to facilitate ethics integration. For example, the Defense Advanced Research Projects Agency (DARPA), one of the federal agencies involved in the BRAIN Initiative, has integrated ethics formally into its neuroscience research efforts. DARPA has convened an independent panel of six nationally recognized bioethicists—not employed by the agency—to inform the ethical conduct of neuroscience.[70] The panel provides expert insight to DARPA program managers on ethical, legal, and social issues associated with their BRAIN Initiative projects. Each program manager consults an ethics mentor at the inception of a project to incorporate ethical considerations according to "the three Cs—character, consent, and consequence."[71] Ethics mentors provide insight on issues such as respecting autonomy through informed consent, or the individual and societal consequences of neurotechnologies.[72] Panel members also write an ethics-focused white paper about each program or project plan. In addition to the services provided by the independent panel, each program is required to set aside funds for similar consultation services or ethics-focused research, as a supplement to the panel's work. In this way, DARPA has combined several approaches to ethics integration— some of which are elaborated in the approaches discussed below.

The Role of Professional Societies

Professional societies can play an important role in fostering a dialogue between the scientific and ethics communities. The Society for Neuroscience and the International Neuroethics Society hold their annual conferences together, ensuring that the location and timing are aligned. This coordination encourages scientists to engage with ethical issues, and ethicists to learn more about emerging developments in neuroscience. In addition, the Society for Neuroscience annual conference features the David Kopf Lecture on Neuroethics. Past topics have included behavioral sciences in the courtroom, the ethics of cognitive enhancement, and managing incidental findings discovered in neuroimaging.

Sources: Massachusetts General Hospital Center for Law, Brain & Behavior. (2013). News and Events: International Neuroethics Society to Meet at SfN 2013 in San Diego. Retrieved April 7, 2014 from http://clbb.mgh.harvard. edu/international-neuroethics-society-tomeet-at-sfn-2013-in-san-diego/; Society for Neuroscience. (2012). Neuroscience 2012 Featured Lectures. Retrieved May 5, 2014, from http://am2012.sfn.org /am2012/events/ lectures.aspx?type=feature&print=on; Illes, J. (2006). Neuroethics, Neurochallenges: A Needs-Based Research Agenda. Retrieved April 7, 2014 from http://neuroethics.stanford. edu/ documents/Illes.NeuroethicsSFN2006.pdf; Society for Neuroscience. (2013). Neuroscience 2013: Featured Lectures. Retrieved April 7, 2014 from http://www.sfn.org /annual-meeting/ neuroscience-2013/abstracts-and-sessions/ scientific-program/featured-lectures.

Ethics Integration through Research about the Ethical, Legal, and Social Implications of Scientific Research

One approach for ethics integration is for funders to direct consideration of the ethical and social implications of the research they support.[73] For example, the National Institutes of Health's (NIH's) and U.S. Department of Energy's Human Genome Project Ethical, Legal, and Social Implications Program (HGP ELSI Program) was the first national model formally established in the United States for integrating ethical and social issues into a major federal science initiative. The agencies involved designated a percentage of their annual extramural budgets to programs examining the ethical, legal, and social ramifications of genetic and genomic research.[74] Setting aside a direct source of funds helped to create parallel and multidisciplinary research on ethical and social aspects of genomics. Previous efforts at science and ethics integration offer important lessons for future funders, scientists, and ethicists. The goals and approaches of the HGP ELSI Program developed over time. Formal review processes, feedback from genetic scientists, and constructive critique from a growing community of interdisciplinary scholars helped inform changes in strategic vision and practical implementation of the Program's research, education, and policy components.[75]

The HGP ELSI Program is one among many ethics integration efforts associated with multinational and country-specific research initiatives (see Appendix I: *Ethics Integration in Selected International Neuroscience Research Programs*).

The Human Microbiome Project Funds Research on Its Ethical, Legal, and Social Implications

The goal of the NIH Human Microbiome Project (HMP) is to identify and characterize the community of microorganisms that live in or on the human body. The HMP currently funds six projects that study the ethical, legal, and social implications of human microbiome research. The projects cover topics such as the concept of risk and benefit in human microbiome research, the current regulatory framework for probiotic products on the market, and the implications of research on ancient and contemporary human microbiomes for the identities of indigenous people.

Sources: Peterson, J., et al. (2009). The NIH HMP. *Genome Research*, 19(12), 2317-2323; NIH HMP. (n.d.). Ethical, legal, and social implications [Webpage]. Retrieved April 11, 2014 from http:// www.hmpdacc.org/ethical/ethical.php.

Ethics Integration through Research Ethics Consultation

Within research institutions, an ethics consultation service is one way to integrate ethics into the research process in neuroscience and other fields. Though not currently commonplace as an institutionalized service, there are established examples of research ethics consultation services.[76]

The NIH Clinical Center (Clinical Center) admits 10,000 new patients a year, all of whom are participants in some form of clinical research.[77] The Clinical Center provides a Bioethics Consultation Service for all NIH staff as well as Clinical Center research participants and families, intended to serve several purposes such as assisting in making decisions about research participation.[78]

Another example of a consultation service is the Benchside Ethics Consultation Service (BECS) at Stanford University, which describes its role as making ethics advice available to bench scientists and clinical researchers "as the science unfolds," influencing study design in its early phases, and helping to ensure that research is not halted or slowed later in the process when ethical issues become more apparent or more problematic.[79] To enhance uptake and effectiveness, BECS implemented measures to increase confidentiality and set realistic expectations about the limits on confidentiality in response to researchers' concerns including "going over the heads of more senior researchers in obtaining a consultation."[80]

Consultation services formalize the longstanding practice of researchers seeking advice from colleagues on an individual basis.[81] There are several advantages to formalizing this practice. First, consultation services comprise teams of ethicists and scientists, who have a range of expertise and bring a variety of perspectives to the consultation service. This composition broadens the scope of considerations that can be brought to bear on an issue. Second, a centralized and formalized process can serve at the institutional level as well as the protocol level, allowing the consultation team to compile data on the types of ethical issues that arise, and to engage institutions, programs, and investigators in a broader conversation about implications of particular types of research. The BECS team has observed that researchers approach the ethicists at various points in the research process, before the initiation of a research project—when investigators are deciding questions of hypothesis, design, and recruitment—during the course of a study, or before or after results are published.

The consultation model allows researchers ongoing access to a team of ethicists, and strives to make ethics a seamless part of the research process.[82]

Ethics Integration through Stakeholder Engagement

Another mechanism for integrating ethics and science is to engage with stakeholders, who often include members of the public, private partners, industry representatives, neuroscientists from many different subfields, patients, research participants, and community members (see Figure 1: Potential Stakeholders in Neuroscience Research). The identification and inclusion of relevant stakeholders is an important component of assessing and addressing ethical issues in specific contexts. Institutions and researchers can use stakeholder engagement strategies to better understand the perspectives and concerns of parties interested in specific research. Approaches to stakeholder engagement include public meetings; public comment; and other community engagement strategies, such as community advisory boards or community-based participatory research. Stakeholder engagement in neuroscience, as in other emerging sciences, is a process that would greatly benefit from enhanced attention by institutions engaged in the field.

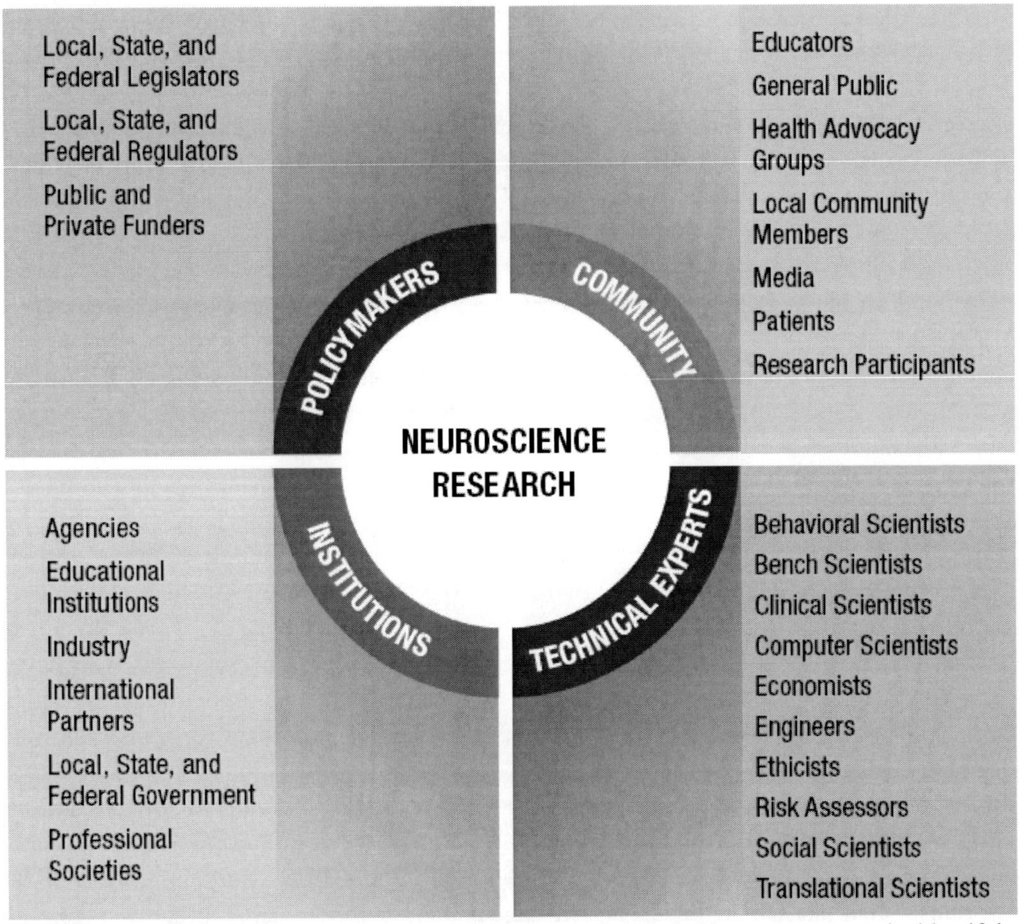

*This list is not exhaustive. An important component of stakeholder engagement is identifying stakeholders through a variety of possible methods including literature review, referral, or community meetings.

Sources: Agency for Healthcare Research and Quality (AHRQ). (2011). *Engaging Stakeholders to Identify and Prioritize Future Research Needs*, (AHRQ Publication No. 11-EHC044-EF). Rockville, MD: AHRQ, p. 7. Retrieved April 7, 2014 from http://www.effectivehealthcare.ahrq. gov/ehc/products/200/698/MFRNGuide04--Engaging_Stakeholders--6-10-2011. pdf; UNAIDS /AVAC. (2011). *Good Participatory Practice: Guidelines for Biomedical HIV Prevention Trials*. Retrieved April 7, 2014 from http://www.avac.org/ht/a/GetDocumentAction/i/35219; MacQueen, K.M., et al. (2012). *Stakeholder Engagement Toolkit for HIV Prevention Trials*. Retrieved April 7, 2014 from http://www.fhi360.org/sites/default/files/media/documents/Stakeholder%20 EngagementToolkit%20for%20HIV%20Prevention%20Trials.pdf.

Figure 1. Potential Stakeholders in Neuroscience Research.*

One example of stakeholder engagement in ethics and science is the National Nanotechnology Initiative's process to generate a 2011 Environmental, Health, and Safety Research Strategy. A nanotechnology working group held a series of four public workshops covering topics that included an ethical, legal, and social implications capstone.[83] The working group obtained information pertinent to setting research priorities and identified gaps and obstacles to accomplishing the needed research by conducting a series of public engagement workshops and seeking public comment. The strategy was published along with a brochure intended to summarize the plan for the public and engaged stakeholders.[84]

Ethics Integration through Inclusion of an Ethics Perspective on the Research Team. Another strategy for integrating ethics and science is to include an ethicist—or a scientist with experience in ethics—directly on the research team or research protocol. This strategy incorporates social science and humanist perspectives into a research project by engaging researchers in critical examination at different decision points in the research process. This approach stimulates awareness of social considerations and incorporates them into research-process decisions (in addition to research-prioritization decisions, for example). Integrating ethical considerations into routine science decision making occurs by embedding ethicists and social scientists into the laboratory where learning is reciprocal. As one researcher who studied the effectiveness of the approach noted, "Rather than experiencing societal considerations as 'ethical speed bumps' imposed on their projects, the [researchers] indicated that such reflections broadened their decisions. They realized that they were making choices, that these choices were based on a range of considerations, and that by reflecting on them, they found that decision outputs and inputs can both vary."[85] Periodic meetings between researchers and persons with experience dealing with ethical issues can be used to discuss researchers' decisions, the relevant considerations, potential alternative choices, and possible outcomes— an approach that can be particularly responsive to the ethical issues facing individual researchers.[86]

* * *

Recognizing and confronting challenges to the integration of ethics and science is an important component of planning and implementation. Adequate funding is necessary (though not sufficient) to achieve effective integration. Inadequate funding presents practical challenges and also can be perceived to reflect a lack of political will for substantive engagement with ethical issues.

Additional potential challenges might include professional incentive structures, available expertise, and the available evidence for particular approaches to integration. First, professional reward and promotion structures, such as tenure or grants, can incentivize or disincentivize early incorporation of ethics into scientific programs.[87] For example, scientists face pressure to run their own laboratories, and are rewarded for conducting their own focused work; as a result, as one neurologist explained to the Bioethics Commission, scientists "don't go out and talk to other people because that's pro bono work.... [W]hat you need to do is advance your science. And our institutions are set up to promote that and not cross disciplinary engagements."[88] Inclusion of professional activities related to integration in professional reward structures such as tenure might encourage researchers to participate. Second, there might be an insufficient supply of available experts with relevant cross-disciplinary fluency or time.[89] The availability of trained professionals will grow with the proliferation of education programs for students and professionals. Finally, there might be gaps in knowledge or evidence about which approach or combination of approaches work best in a given context. Just as previous ethics integration efforts, such as the HGP ELSI Program, offered important lessons, increasing implementation and evaluation of various approaches to neuroscience ethics integration will advance the scholarship and body of evidence over time.[90]

It is important to acknowledge the practical, professional, and political challenges that those undertaking ethics integration sometimes face. To enhance the likelihood of success, there should be a clear vision of, and a strategic approach to, ethics integration to help identify and overcome obstacles. Funding is critical. It must be accompanied by investment in developing expertise and thoughtful articulation of criteria for success. The ethics integration approaches described here can initiate thoughtful planning that is responsive to a variety of research contexts, from basic science to clinical research. The Bioethics Commission's recommendations are intended to strengthen ethics integration infrastructure for the BRAIN Initiative and neuroscience more broadly, ensuring that this important science will continue to advance the public good.

Recommendations[91]

Ethical issues arise throughout the scientific research process. Those involved in research at all levels, including science policymakers, public and private funding institutions, and individual investigators, have a professional obligation to identify and address ethical issues encountered in their work and a broader obligation to support scientific research that furthers the public good. Fulfillment of these obligations supports scientific quality and is crucial to maintaining public trust essential for scientific progress.

Ethics integration should aid scientists in recognizing and addressing ethical issues as they arise, and ethicists in understanding the science and technology with which they engage. The goal is not to make scientists into ethicists or vice versa. It is to cultivate sufficient fluency for productive discussion and collaboration in a multidisciplinary endeavor that can include biological scientists, clinicians, historians, lawyers, philosophers, physical scientists, social scientists, and theologians, among others. Although ethics is already integrated into science in various ways—as it is in many human activities—more explicit and systematic integration serves to elucidate implicit ethical judgments and allows their merits to be assessed thoughtfully. Moreover, integration efforts can serve both to address ethical issues in particular cases as well as to cultivate an ethical sensibility beyond the bounds of the case at hand.[92]

Recommendation 1:
Integrate Ethics Early and Explicitly Throughout Research

Institutions and individuals engaged in neuroscience research should integrate ethics across the life of a research endeavor, identifying the key ethical questions associated with their research and taking immediate steps to make explicit their systems for addressing those questions. Sufficient resources should be dedicated to support ethics integration.

Existing approaches to ethics integration, as described on previous pages, include but are not limited to educational support, funding mechanisms, ethics consultation, stakeholder engagement, and incorporation of ethics expertise into research teams. Depending on their structure, research priorities, and capacity (among other things), institutions and individuals will implement different systems to integrate ethics and neuroscience research. For some, one approach to integration will be sufficient; for others, combined approaches might be necessary. Yet, in all cases, ethics integration should be explicitly acknowledged and supported to highlight the importance of robust ethical reflection and avoid formulaic approaches that might mask underlying complexity.

When planning for and conducting neuroscience research, it is important that researchers prepare for ethical issues that might arise. When responding to funding opportunities, for example, neuroscience researchers might seek funding for ethics integration. Researchers also might seek support to develop innovative approaches to ethics integration and evaluate approaches' effectiveness. Institutions, such as universities, neuroscience departments, and ethics programs, should seek to integrate ethics on a broader scale. They might develop means to integrate ethics into neuroscience research, for example, through curriculum development or broad stakeholder engagement. These institutions also might seek funding to support ethics integration and evaluation. Research funders should take action as well, such as by requiring ethical expertise as part of research teams, supporting ethics integration within neuroscience research projects, or funding innovation and evaluation of integration in practice (as discussed further in Recommendation 2).

Without dedicated resources that demonstrate commitment to addressing ethical and social issues, successful integration is impossible. Available resources must include financial resources, human capital, and expertise. Funding alone is necessary but not sufficient. Particular attention should be paid to developing and supporting a well-trained cadre of professionals. Resources allocated for integration support research; they do not compete with it. Potential funding sources include public and private institutions involved or with a stake in neuroscience research such as government agencies, private foundations, and universities. Institutions both affirm their commitment to ethics integration and enable such integration to occur through sufficient investment to assess integration needs, implement ethics integration, and make explicit existing systems of integration.

Once implemented, systems for integrating ethics and neuroscience research should not operate in a vacuum. Institutions and individuals engaged in neuroscience research should learn from collective experience to improve existing systems and inform others about what works and what does not.

Recommendation 2:
Evaluate Existing and Innovative Approaches to Ethics Integration

Government agencies and other research funders should initiate and support research that evaluates existing as well as innovative approaches to ethics integration. Institutions and individuals engaged in neuroscience research should take into account the best available evidence for what works when implementing, modifying, or improving systems for ethics integration.

Evaluating approaches to ethics integration goes hand-in-hand with developing innovative approaches. Innovation prevents stagnation and counteracts reliance on approaches that are familiar, but potentially outdated, inefficient, or counterproductive. Evaluation informs innovation, allows for continuous improvement, and measures success. Both development and evaluation contribute to an iterative process toward better ethics integration in neuroscience research.

In evaluating their own systems and other approaches to ethics integration, funders, institutions, and researchers should look to best practices in program evaluation, such as process and outcome evaluation, that are most relevant. Key questions will include how to define criteria for success and how best to measure them. Researchers should share widely their findings and conclusions to support more robust and informed ethics integration.

One foundational approach to integration is pairing science and ethics education at all levels of education. The Bioethics Commission is committed to doing its part to enhance bioethics education. Ethics education takes place at many stages throughout the moral development of individuals, from informal lessons from parents to professional development later in one's career. Early ethics education in academic settings is critical to prepare future scientists to integrate ethical considerations into their work—including future research in neuroscience. Professional development for experienced investigators is equally important and can serve multiple ends, contributing not only to their individual knowledge, but to the knowledge of the students and young scientists that they mentor as well. Ethics education has a better chance of informing action when it is continually reinforced and connected to practical experience.

The Bioethics Commission heard from many scientists and educators that it is essential to ensure that all scientists understand the role of ethics in good science. Graduate school is the most specialized stage of education and bioethics education should continue through all stages of a scientist's education, as ethical issues often become more complex as one advances. But graduate school is certainly not the best place to begin; foundations for ethical thinking must be laid earlier, at all levels. Moreover, these foundations should be continually reinforced—from pre-college education through continuing education for experienced scientists.

The ethical challenges that neuroscience raises are not wholly unique, and the ethics needed to ensure its progress overlap with the ethics needed to ensure progress in all science. Undergraduate liberal arts education in ethics, professional ethics, and the ethics of science—and earlier foundational education at the primary and secondary levels—will better prepare future researchers than starting foundational ethics education at the professional level. Repeated calls for graduate-level approaches are not only impractical but potentially impossible to execute effectively. It is important that ethics education take place at many points before, during, and after graduate school. Continuing professional education after graduate school will reinforce ethical perspectives, building on the strong foundation laid earlier in a scientist's education and career.

Recommendation 3:

Integrate Ethics and Science through Education at All Levels

Government agencies and other research funders should initiate and support research that develops innovative models and evaluates existing and new models for integrating ethics and science through education at all levels.

Some ethics education currently occurs at the secondary and undergraduate levels. For example, NIH offers a bioethics curriculum for high school students.[93] A pilot program testing the curriculum's effectiveness found that learning about bioethics fostered students' curiosity in the underlying science. Science fairs or pre-college competitions such as the Intel Science Talent Search or the For Inspiration and Recognition of Science and Training (FIRST) Robotics Competition could also provide opportunities for instilling the importance of ethics integration as young people contemplate a science or technology career.[94] Some continuing ethics education is occurring as well, for example, through mentoring programs for scientists at all levels, such as the Society for Neuroscience online mentoring community.[95]

While current efforts are promising, further development and evaluation of models, including curricula, for integrating ethics and science through education at all levels are needed. Innovative methods to develop critical thinking, ethical sensitivity, and moral reasoning will provide a strong foundation for students who might later pursue science as a profession, and build on that foundation for more experienced scientists. Models should be widely shared for implementation and evaluation. As with evaluating other approaches to ethics integration, researchers should look to best practices in program evaluation and should share openly their results and findings. Informed by the best available evidence, successful education models should be implemented to better prepare future scientists and engage practicing scientists—including neuroscientists—to integrate ethics into their research.

It has been just over a year since the announcement of the BRAIN Initiative, and institutions participating in this research effort have an important opportunity to integrate ethics and science from the outset.[96] A key component of this integration is the inclusion of ethicists or scientists with experience in ethics in BRAIN Initiative-related scientific advisory boards and funding review committees, particularly for the major public and private sector partners.

Recommendation 4:

Explicitly Include Ethical Perspectives on Advisory and Review Bodies

BRAIN Initiative-related scientific advisory and funding review bodies should include substantive participation by persons with relevant expertise in the ethical and societal implications of the neuroscience research under consideration.

Major public and private sector partners might achieve this goal in different ways, aligned with their organizational structure and function. For example, DARPA's approach relies on an independent neuroscience ethics advisory panel. NIH or the National Science Foundation might choose to include ethicists on relevant scientific advisory and planning groups or study sections reviewing responses to BRAIN Initiative funding announcements. Ethical perspectives should be included appropriately on neuroscience research advisory bodies beyond the BRAIN Initiative as well.

No matter how institutions approach this aspect of ethics integration, they should ensure that scientific governance in neuroscience is imbued with an ethical perspective. For this to be successful, the mandates, structures, and processes of relevant advisory bodies should ensure that ethicists are given the opportunity to make meaningful contributions to discussions and decisions capable of influencing the research endeavor as a whole.[97]

* * *

Ethics is an important companion to science that both facilitates and enhances the scientific endeavor. Neuroscience research in general and the BRAIN Initiative in particular present important opportunities to integrate science and ethics for their mutual benefit. Successful ethics integration requires commitment, innovation, sensitivity to context, and adequate resources.

End Notes

[1] Letter from President Barack Obama to Dr. Amy Gutmann, Chair, Presidential Commission for the Study of Bioethical Issues (PCSBI). (2013, July 1). Retrieved April 4, 2014 from http://bioethics.gov/sites/default/files/news/Charge%20from%20President%20Obama.pdf; Obama, B. (2013, April 2). Remarks by the President on the BRAIN Initiative and American Innovation. Retrieved April 4, 2014 from http://www.whitehouse.gov/the-press-office/2013/04/02/remarks-president-brain-initiative-and-american-innovation.

[2] Building on its past work on neuroscience and related ethical issues, including hearing from experts at two public meetings, the Bioethics Commission held three public meetings and heard from representatives from public agencies and private sector partners involved in the BRAIN Initiative, neuroscientists, philosophers, and other experts discussing the importance of and approaches to integrating ethics and science. The Bioethics Commission also published a request for public comment in the *Federal Register* and received several thoughtful and insightful responses.

[3] Ramón y Cajal, S. (1956). *Charlas de café: Pensamientos, Anécdotas, y Confidencias*, Séptima Edición [Coffee Chat: Thoughts, Anecdotes, and Confidences, Seventh Edition]. Madrid: Espasa-Calpe, S.A, p. 181. ('El cerebro humano representa un muncho donde figuran algunos continentes explorados y vastas tierras ignotas." Translated by Presidential Commission for the Study of Bioethical Issues Staff.).

[4] World Health Organization (WHO). (2006). *Neurological Disorders: Public Health Challenges*, Chapter 4: Conclusions and Recommendations, p. 177. Retrieved April 4, 2014 from http://www.who.int/mental_health/neurology/chapter_4_neuro_disorders_public_h_challenges.pdf; Hirtz, D., et al. (2007). How common are the 'common" neurologic disorders? *Neurology*, 68(5), 326-337; Insel, T.R., and S.C. Landis. (2013). Twenty-five years of progress: The view from NIMH and NINDS. *Neuron*, 80(3), 561-567.

[5] See, e.g., Capron, A.M., Board Chair, Public Responsibility in Medicine and Research (PRIM&R). (2014, March 26). Comments submitted to PCSBI ('The ethics of neuroscience research is not distinct from the ethics of research in other areas of biology and medicine.").

[6] Safire, W. (2002). Visions for a new field of 'neuroethics." In Sj. Marcus (Ed.). *Neuroethics: Mapping the Field* (pp. 3-9). New York, NY: The Dana Foundation.

[7] Caulfield, T., Canada Research Chair in Health Law and Policy, and Professor in the Faculty of Law and the School of Public Health, University of Alberta. (2014). Science 'Hype": Sources and Solutions. Presentation to PCSBI, February 10. Retrieved April 4, 2014 from http://bioethics.gov/sites/default/files/Caulfield%20 Meeting%2016%20Presentation.pdf; Neill, U., Director, Office of the President, Memorial Sloan-Kettering Cancer Center; and Editor at Large, *Journal of Clinical Investigation.* (2014). Ethics of Communication about Neuroscience Research by Scientists and Journalists. Presentation to PCSBI, February 10. Retrieved April 4, 2014 from http://bioethics.gov/sites/default/files/Neill%20Meeting%2016%20Presentation.pdf; Racine, E., Director, Neuroethics Research Unit; and Associate Research Professor, Institut de Recherches Cliniques de Montréal; Associate Research Professor, Department of Medicine, Université de Montréal; Adjunct Professor, Department of Medicine and Department of Neurology and Neurosurgery, McGill University. (2014). Neuroscientists and the Public: Perspectives and Questions from Neuroethics. Presentation to PCSBI, February 10. Retrieved April 4, 2014 from http://bioethics.gov/sites/default/files/Racine%20Meeting%20 16%20Presentation.pdf; Ward, Sj.A., Professor and Director, George S. Turnbull Center, School of Journalism and Communication, University of Oregon-Portland. (2014). Journalism Ethics Amid a Media Revolution. Presentation to PCSBI, February 10. Retrieved April 4, 2014 from http://bioethics.gov/sites/default/files/ Ward%20Meeting%2016%20Presentation.pdf.

[8] Dumit, J. (2004). *Picturing Personhood: Brain Scans and Biomedical Identity*. Princeton, NJ: Princeton University Press; Farah, Mj. (2005). Neuroethics: The practical and the philosophical. *Trends in Cognitive Sciences*, 9(1), 34-40; Giordano, J., Kulkarni, A., and J. Farwell. (2014). Deliver us from evil? The temptations, realities, and neuroethico-legal issues of employing assessment neurotechnologies in public safety initiatives. *Theoretical Medicine and Bioethics*, 35(1), 73-89; Santosh, Pj. (2000). Neuroimaging in child and adolescent psychiatric disorders. *Archives of Disease in Childhood*, 82(5), 412-419.

[9] Farah, Mj., et al. (2014). Functional MRI-based lie detection: Scientific and societal challenges. *Nature Reviews Neuroscience*, 15(2), 123-131; Jones, O.D., et al. (2013). Law and neuroscience. *The Journal of Neuroscience*, 33(45), 17624-17630.

[10] Farah, Mj., *supra* note 9; Wolpe, P.R., Foster, K.R., and D.D. Langleben. (2010). Emerging neurotechnologies for lie-detection: Promises and perils. *American Journal of Bioethics*, 10(10), 40-48; Langleben, D.D., et al. (2002). Brain activity during simulated deception: An event-related functional magnetic resonance study. *Neuroimage*, 15(3), 727-732; Kozel, F.A., et al. (2005). Detecting deception using functional magnetic resonance imaging. *Biological Psychiatry*, 58(8), 605-613; Langleben, D.D., et al. (2005). Telling truth from lie in individual subjects with fast event-related fMRI. *Human Brain Mapping*, 26(4), 262-272. Although some scholars believe that fMRI will replace traditional polygraph in certain settings, others caution that, "although brain imaging is a more direct index of cognition than the traditional polygraph, it is subject to many of the same caveats." Monteleone, G.T., et al. (2009). Detection of deception using fMRI: Better than chance, but well below perfection. *Social Neuroscience*, 4(6), 528-538, p. 528.

[11] Kozel, F.A., *supra* note 10.

[12] Ibid, p. 612.

[13] Moyer, V.A. (2014). Screening for cognitive impairment in older adults: U.S. Preventive Services Task Force recommendation statement. *Annals of Internal Medicine*, doi:10.7326/M14-0496. Alzheimer's disease is the most common form of dementia. WHO. (2012). Dementia [Webpage]. Retrieved April 5, 2014 from http://www. who.int/mediacentre/factsheets/fs362/en/. A recent estimate suggests that over 500,000 people over aged 75 died of Alzheimer's disease in the United States in 2010. James, B.D., et al. (2014). Contribution of Alzheimer disease to mortality in the United States. *Neurology*, 82(12), 1045-1050.

[14] Centers for Disease Control and Prevention (CDC). Dementia/Alzheimer's Disease [Webpage]. Retrieved April 9, 2014 from http://www.cdc.gov/mentalhealth/basics/mental-illness/dementia.htm.

[15] Moyer, V.A., *supra* note 13.

[16] Leuzy, A., and S. Gauthier. (2012). Ethical issues in Alzheimer's disease: An overview. *Expert Review of Neurotherapeutics*, 12(5), 557-567.

[17] Mapstone, M., et al. (2014). Plasma phospholipids identify antecedent memory impairment in older adults. *Nature Medicine*, doi:10.1038/nm.3466; Leuzy, A., and S. Gauthier, *supra* note 16; Mattsson, N., Brax, D.,

and H. Zetterberg. (2010). To know or not to know: Ethical issues related to early diagnosis of Alzheimer's disease. *International Journal of Alzheimer's Disease*, 2010, 1-4.

[18] Buchanan, A. (1988). Advance directives and the personal identity problem. *Philosophy & Public Affairs*, 17(4), 277-302; Dresser, R. (1986). Life, death, and incompetent patients: Conceptual infirmities and hidden values in the law. *Arizona Law Review*, 28(373), 373-405; Dresser, R.S. (1989). Advance Directives, Self-determination, and Personal Identity. In C. Hackler, R. Moseley, and D.E. Vawter (Eds.). *Advance Directives in Medicine* (pp. 155-170). New York, NY: Praeger; Dworkin, R. (1994). *Life's Dominion: An Argument About Abortion, Euthanasia, and Individual Freedom*. New York, NY: Vintage Books; Jaworska, A. (1999). Respecting the margins of agency: Alzheimer's patients and the capacity to value. *Philosophy and Public Affairs*, 28(2), 105-138.

[19] Dworkin, R., *supra* note 18, pp. 222-229.

[20] Dresser, R.S. (1989), *supra* note 18; Jaworska, A., *supra* note 18.

[21] Beauchamp, T.L., and J.F. Childress. (2012). *Principles of Biomedical Ethics*, Seventh Edition. Oxford: Oxford University Press.

[22] Dresler, M., et al. (2013). Non-pharmacological cognitive enhancement. *Neuropharmacology*, 64, 529-543.

[23] Butcher, J. (2003). Cognitive enhancement raises ethical concerns: Academics urge pre-emptive debate on neurotechnologies. *Lancet*, 362(9378), 132-133; Greely, H., et al. (2008). Towards responsible use of cognitive-enhancing drugs by the healthy. *Nature*, 456(7223), 702-705.

[24] Greely, H., et al., *supra* note 23.

[25] Daniels, N. (2008). *Just Health: Meeting Health Needs Fairly*. New York, NY: Cambridge University Press, pp. 149-155; Buchanan, A., et al. (2000). *From Chance to Choice: Genetics and Justice*. New York, NY: Cambridge University Press.

[26] Chatterjee, A. (2006). The promise and predicament of cosmetic neurology. *Journal of Medical Ethics*, 32(2), 110-113; Greely, H., et al., *supra* note 23.

[27] Dietz, P., et al. (2013). Randomized response estimates for the 12-month prevalence of cognitive-enhancing drug use in university students. *Pharmacotherapy*, 33(1), 44-50; Wilens, T. (2008). Misuse and diversion of stimulants prescribed for ADHD: A systematic review of the literature. *Journal of the American Academy of Child and Adolescent Psychiatry*, 47(1), 21-31; Teter, C.J., et al. (2006). Illicit use of specific prescription stimulants among college students: Prevalence, motives, and routes of administration. *Pharmacotherapy*, 26(10), 1501-1510; Smith, M.E., and M.J. Farah. (2011). Are prescription stimulants "smart pills"? The epidemiology and cognitive neuroscience of prescription stimulant use by normal healthy individuals. *Psychological Bulletin*, 137(5), 717-741.

[28] Emanuel, R.M., et al. (2013). Cognitive enhancement drug use among future physicians: Findings from a multi-institutional census of medical students. *Journal of General Internal Medicine*, 28(8), 1028-1034.

[29] Ibid, p. 1032.

[30] American College Health Association. (2013). *American College Health Association-National College Health Assessment II: Undergraduate Students-Reference Group Executive Summary Spring 2013*. Hanover, MD: American College Health Association.

[31] McCabe, S.E., et al. (2005). Non-medical use of prescription stimulants among US college students: Prevalence and correlates from a national survey. *Addiction*, 99(1), 96-106, p. 98.

[32] Ibid, pp. 96-106.

[33] Substance Abuse and Mental Health Services Administration, Office of Applied Studies. (2009). *The NSDUH Report: Nonmedical Use of Adderall Among Full-Time College Students*. Retrieved April 17, 2014 from http://www.samhsa.gov/data/2k9/adderall/adderall.pdf.

[34] Chatterjee, A., *supra* note 26.

[35] Collins, K.L., et al. (2010). Deep brain stimulation for movement disorders. *Neurobiology of Disease*, 38(3), 338-345. Although the benefits of using DBS to treat motor symptoms of Parkinson's disease are considerable, research has not clearly shown that DBS alters the long-term progression of non-motor symptoms of Parkinson's disease, including impaired cognition, sleep disturbances, and autonomic dysfunction. See Borgohain, R., et al. (2012). Nonmotor outcomes in Parkinson's disease: Is deep brain stimulation better than dopamine replacement therapy? *Therapeutic Advances in Neurological Disorders*, 5(1), 23-41.

[36] Clausen, J. (2010). Ethical brain stimulation—Neuroethics of deep brain stimulation in research and clinical practice. *European Journal of Neuroscience*, 32(7), 1152-1162.

[37] Johnson, J. (2009). A dark history: Memories of lobotomy in the new era of psychosurgery. *Medicine Studies*, 1(4), 367-378; National Commission for the Protection of Human Subjects of Biomedical and Behavioral

Research. (1977). *Report and Recommendations: Psychosurgery*. Washington, DC: Department of Health, Education, and Welfare (DHEW).

[38] Feldman, R.P., and J.T. Goodrich. (2001). Psychosurgery: A historical overview. *Neurosurgery*, 48(3), 647-659; Rosenfeld, J.V., and J.H. Lloyd. (1999). Contemporary psychosurgery. *Journal of Clinical Neuroscience*, 6(2), 106-112.

[39] Feldman, R.P., and J.T. Goodrich, *supra* note 38; Rosenfeld, J.V., and J.H. Lloyd, *supra* note 38; Lerner, B.H. (2005). Last-ditch medical therapy—Revisiting lobotomy. *New England Journal of Medicine*, 353(2), 120-121.

[40] Feldman, R.P., and J.T. Goodrich, *supra* note 38; Rosenfeld, J.V., and J.H. Lloyd, *supra* note 38.

[41] Feldman, R.P., and J.T. Goodrich, *supra* note 38; Diefenbach, G.J., et al. (1999). Portrayal of lobotomy in the popular press: 1935-1960. *Journal of the History of Neurosciences*, 8(1), 60-69.

[42] Feldman, R.P., and J.T. Goodrich, *supra* note 38, p. 655.

[43] Lerner, B.H., *supra* note 39.

[44] Rabins, P., et al. (2009). Scientific and ethical issues related to deep brain stimulation for disorders of mood, behavior and thought. *Archives of General Psychiatry*, 66(9), 931-937; Schlaepfer, T.E., George, M.S., and H. Mayberg. (2010). WFSBP guidelines on brain stimulation treatments in psychiatry. *The World Journal of Biological Psychiatry*, 11(1), 2-18.

[45] Rabins, P., et al., *supra* note 44.

[46] Ibid; Presidential Commission for the Study of Bioethical Issues. (2013, December). *Anticipate and Communicate: Ethical Management of Incidental and Secondary Findings in the Clinical, Research, and Direct-toConsumer Contexts*. Washington, DC: PCSBI, pp. 78-79.

[47] Clausen, J., *supra* note 36; Patterson, T. (2014, January 13). My brain implant saved me from depression. *The Huffington Post*. Retrieved April 7, 2014 from http://www.huffingtonpost.com/teerika-patterson/depression_b_4559861.html.

[48] Lin, J.M., et al. (2008). An analysis of the abstracts presented at the annual meetings of the Society for Neuroscience from 2001 to 2006. *PLoS ONE*, 3(4), e2052; Lemm, S., et al., (2010). Introduction to machine learning for brain imaging. *Neuroimage*, 56(2), 387-399; Deisseroth, K., and M.J. Schnitzer. (2013). Engineering approaches to illuminating brain structure and dynamics. *Neuron*, 80(3), 568-577; Amari, S. (2013). Dreaming of mathematical neuroscience for half a century. *Neural Networks*, 37, 48-51; Minati, L., et al. (2013). From brain topography to brain topology: Relevance of graph theory to functional neuroscience. *NeuroReport*, 24(10), 536-543.

[49] Potter, V.R. (1971). *Bioethics: Bridge to the Future*. Englewood Cliffs, NJ: Prentice-Hall, Inc.

[50] Solomon, M.Z., President, The Hastings Center; and Clinical Professor of Anaesthesia, Harvard Medical School. (2013). Transformational Bioethics Learning in Brain Science. Presentation to PCSBI, December 18. Retrieved April 7, 2014 from http://bioethics.gov/node/3238.

[51] Ibid.

[52] Presidential Commission for the Study of Bioethical Issues. (2014). Implementation Strategies for Ethics Integration Roundtable Discussion with Members and Presenters, February 11. Retrieved April 7, 2014 from http://bioethics.gov/node/3379.

[53] Khurana, V., and S. Lindquist. (2010). Modelling neurodegeneration in *Saccharomyces cerevisiae*: Why cook with baker's yeast? *Nature Reviews Neuroscience*, 11(6), 436-449; Steneck, N.H. (2006). Fostering integrity in research: Definitions, current knowledge, and future directions. *Science and Engineering Ethics*, 12(1), 53-74; Kovac, J. (2006). Professional ethics in science. *Boston Studies in the Philosophy of Science*, 242, 157-169.

[54] Douglas, H.E. (2009). *Science, Policy, and the Value-Free Ideal*. Pittsburgh, PA: University of Pittsburgh Press, p. 72.

[55] Harris, J. (2005). Scientific research is a moral duty. *Journal of Medical Ethics,* 31(4), 242-248.

[56] Koepsell, D. (2010). On genies and bottles: Scientists' moral responsibilities and dangerous technology R&D. *Science and Engineering Ethics*, 16(1), 119-133; Douglas, H.E., *supra* note 54, pp. i-210; Kitcher, P. (2001). *Science, Truth, and Democracy*. New York, NY: Oxford University Press, Chapter 10.

[57] Presidential Commission for the Study of Bioethical Issues. (2010, December). *New Directions: The Ethics of Synthetic Biology and Emerging Technologies*. Washington, DC: PCSBI, pp. 24-32.

[58] Lombardo, P.A. (2008). *Three Generations, No Imbeciles: Eugenics, the Supreme Court, and Buck v. Bell.* Baltimore: Johns Hopkins University Press; Kevles, D.J. (1985). *In the Name of Eugenics: Genetics and the Uses of Human Heredity*. New York, NY: Knopf.

[59] E.g., Giordano, J., Chief Neuroethics Studies Program, Center for Clinical Bioethics and Division of Integrative Physiology, Department of Biochemistry, Georgetown University Medical Center. (2014, February 5).

Comments submitted to PCSBI (commenting on the importance of educating both professionals and the general public about neuroscience and neurotechnology).

[60] Eisen, A., and R.M. Berry. (2002). The absent professor: Why we don't teach research ethics and what to do about it. *American Journal of Bioethics*, 2(4), 38-49.

[61] Integration of ethics through education at all levels is especially important in light of evidence that suggests that some current models of ethics education at the graduate level are ineffective. Shorr, A.F., Hayes, R.P., and J.F. Finnerty. (1994). The effect of a class in medical ethics on first-year medical students. *Academic Medicine*, 69(12), 998-1000; Antes, A.L., et al. (2010). Evaluating the effects that existing instruction on responsible conduct of research has on ethical decision making. *Academic Medicine*, 85(3), 519-526.

[62] Wolpe, P.R., Director, Center for Ethics, Asa Griggs Candler Professor of Bioethics, Emory University. (2013). Integrating Ethics and Neuroscience through Education. Presentation to PCSBI, December 18. Retrieved April 7, 2014 from http://bioethics.gov/node/3238.

[63] Solomon, M.Z., *supra* note 50.

[64] Ibid.

[65] Solomon, M.Z., *supra* note 50; National Institutes of Health (NIH). (2009). Exploring Bioethics. Retrieved April 4, 2014 from http://science.education.nih.gov/supplements/nih9/bioethics/guide/pdf/teachers_guide.pdf.

[66] The Dana Foundation. (2014). Brain Awareness Week [Webpage]. Retrieved April 4, 2014 from http://www.dana.org/BAW/; The Dana Foundation. (2014). Brain Awareness Week Calendar of Events [Webpage]. Retrieved April 4, 2014 http://www.dana.org/baw/calendar/; Society for Neuroscience. (2014). Brain Awareness Campaign [Webpage]. Retrieved April 4, 2014 from http://www.sfn.org/public-outreach/brain-awareness-week.

[67] Duke Science & Society. (2014). Concentrations [Webpage]. Retrieved April 4, 2014 from http://scienceandsociety.duke.edu/learn/ma/curriculum/concentrations/; Duke Science & Society. (2014). FAQs [Webpage]. Retrieved April 4, 2014 from http://scienceandsociety.duke.edu/learn/ma/curriculum/faqs/.

[68] Center for Neuroscience & Society. (2013). Teaching Resources [Webpage]. Retrieved April 4, 2014 from http://neuroethics.upenn.edu/?page_id=732.

[69] Research Councils UK. (2014). Interdisciplinary Network on Teaching of Ethics for Neuroscientists [Webpage]. Retrieved April 4, 2014 from http://gtr.rcuk.ac.uk/project/8DD27A6D-FAF9-40AF-9411- A3D325990D42; The University of Manchester. (n.d.). Neuroethics Education: Interdisciplinary Network on Teaching of Ethics for Neuroscientists. Retrieved April 4, 2014 from http://www.lab.ls.manchester.ac.uk/ neuroethicseducation/.

[70] Underwood, E. (2013, November 28). DARPA wants to fix broken brains, restore lost memories. *Science Insider*, Retrieved April 4, 2014 from http://news.sciencemag.org/brain-behavior/2013/11/ darpa-wants-fix-broken-brains-restore-lost-memories.

[71] Casebeer, W.D., Program Manager, Defense Advanced Research Projects Agency (DARPA). (2013). Ethical Issues Associated with the BRAIN Initiative and Ongoing Work in Neuroscience. Presentation to PCSBI, August 20. Retrieved April 4, 2014 from http://bioethics.gov/node/2779.

[72] Ibid.

[73] Funders also could support foundational research in philosophy, theology, history, and social science, among others.

[74] Cook-Deegan, R.M. (1994). *The Gene Wars: Science, Politics, and the Human Genome*. New York, NY: W.W. Norton. Retrieved April 4, 2014 from http://www.genome.duke.edu/press/books/gene-wars/; ELSI Research Planning and Evaluation Group (ERPEG). (2000). *A Review and Analysis of the Ethical, Legal, and Social Implications (ELSI) Research Programs at the National Institutes of Health and the Department of Energy: Final Report of the ELSI Research Planning and Evaluation Group*. Retrieved April 7, 2014 from http://www.genome. gov/10001727.

[75] ERPEG. (2000, February 10). Review and Analysis of the ELSI Research Programs at the National Institutes of Health and the Department of Energy (ERPEG Final Report). Retrieved from http://www.genome.gov/Pages/Research/DER/ELSI/erpeg_report.pdf; National Human Genome Research Institute (NHGRI). (2002). The Role of ELSI Research and Policy Activities in the NHGRI Plan (The Role Report) [Webpage]. Retrieved April 21, 2014 from http://www.genome.gov/10005516; ELSI Research Advisors (ERA). (2005). *ELSI Research Advisors Report to the National Advisory Council for Human Genome Research*. Retrieved April 21, 2014 from http://www.genome.gov/pages/research/der/ELSI/ERA_ELSI_Report_Feb_2005.pdf; Jeungst, E.T. (1996). Self-critical federal science? The ethics experiment within the U.S. Human Genome Project. *Social Philosophy and Policy*, 13(2), 63-95; Yesley, M.S. (2008). What's ELSI got to do with it? Bioethics and the Human Genome Project. *New Genetics and Society*, 27(1), 1-6; Caulfield, T., et al. (2013). Harm, hype, and

evidence: ELSI research and policy guidance. *Genome Medicine*, 5(3), 21; Lindee, M.S. (1994). The ELSI hypothesis. *Isis*, 85(2), 293-296.

[76] Cho, M.K., et al. (2008). Research ethics consultation: The Stanford experience. *IRB: Ethics & Human Research*, 30(6), 1-6.

[77] NIH Clinical Center. (2013). NIH Clinical Center: There's No Other Hospital Like It [Webpage]. Retrieved April 7, 2014 from http://clinicalcenter.nih.gov/ccc/crc/index.html.

[78] NIH Clinical Center, Department of Bioethics. (n.d.). Clinical and Research Ethics Services: Consult Service [Webpage]. Retrieved April 7, 2014 from http://www.bioethics.nih.gov/clinical/; NIH Clinical Center. (n.d.). Bioethics Consultation Service at the NIH. Retrieved April 7, 2014 from http://www.bioethics.nih.gov /clinical/ brochure.pdf.

[79] Cho, M.K., et al. (2008). Strangers at the benchside: Research ethics consultation. *American Journal of Bioethics*, 8(3), 4-13, p. 4.

[80] Cho, M.K., et al., *supra* note 76, p. 2.

[81] Cho, M.K., et al., *supra* note 76; McCormick, J.B., et al. (2013). The establishment of research ethics consultation services (RECS): An emerging research resource. *Clinical and Translational Science*, 6(1), 40-44.

[82] Cho, M.K., et al. *supra* note 79; Cho, M.K., et al., *supra* note 76.

[83] National Nanotechnology Coordination Office. (2011). *National Nanotechnology Initiative: Environmental, Health, and Safety Research Strategy*, pp. 5-6. October 20. Retrieved April 7, 2014 from http://www.nano.gov/ sites/default/files/pub_resource/nni_2011_ehs_research_strategy.pdf.

[84] Ibid; National Nanotechnology Coordination Office. (2011). Key Concepts in the 2011 National Nanotechnology Initiative: Environmental, Health, and Safety Research Strategy. Retrieved April 7, 2014 from http://www.nano.gov/sites/default/files/pub_resource/nni_2011_ehs_research_strategy.pdf.

[85] Schuurbiers, D., and E. Fisher. (2009). Lab-scale intervention. *European Molecular Biology Organization Reports*, 10(5), 424-427, p. 425.

[86] van Gorp, A., and S. van der Molen. (2011). Parallel, embedded or just part of the team: Ethicists cooperating within a European security research project. *Science and Engineering Ethics*, 17(1), 31-43.

[87] Kahn, J. (2011). The two (institutional) cultures: A consideration of structural barriers to interdisciplinarity. *Perspectives in Biology and Medicine*, 54(3), 399-408.

[88] Chatterjee, A., Professor of Neurology, Center for Cognitive Neuroscience and Center for Functional Neuroimaging, University of Pennsylvania School of Medicine. (2013). Roundtable Discussion. Presentation to PCSBI, August 20. Retrieved April 7, 2014 from http://bioethics.gov/node/2781.

[89] ERPEG, *supra* note 74.

[90] ERPEG, *supra* note 74; National Advisory Council for Human Genome Research (NACHGR). (2008). *ELSI Assessment Panel (EAP) Report*. Retrieved April 7, 2014 from http://www.genome.gov/Pages/About /NACHGR/EAPReportFinal.pdf.

[91] These recommendations are summarized in Appendix II: *Summary of Bioethics Commission Recommendations*.

[92] de Melo-Martín, I. (2009). Creating reflective spaces: Interactions between philosophers and biomedical scientists. *Perspectives in Biology and Medicine*, 52(1), 39-47.

[93] NIH, *supra* note 65.

[94] For Inspiration and Recognition of Science and Technology (FIRST). (n.d.). Vision and Mission [Webpage]. Retrieved April 7, 2014 from http://www.usfirst.org/aboutus/vision; Intel Science Talent Search. (n.d.). Intel STS Home [Webpage]. Retrieved April 7, 2014 from https://student.societyforscience.org/intel-sts.

[95] Society for Neuroscience. (n.d.). Mentoring and Career Advice. [Webpage]. Retrieved April 7, 2014 from http:// www.sfn.org/mentoring.

[96] Obama, B. (2013, April 2), *supra* note 1.

[97] Previous efforts such as the HGP ELSI program can provide lessons about the importance and role of multidisciplinary science advisory processes. ERPEG, *supra* note 74; NACHGR, *supra* note 90, pp. 7-8. Lessons also might be drawn from institutional review board (IRB) members' experiences working with community members. Sengupta, S., and B. Lo. (2003). The roles and experiences of nonaffiliated and non-scientist members of institutional review boards. *Academic Medicine*, 78(2), 212-218.

APPENDICES

Appendix I: Ethics Integration in Selected International Neuroscience Research Programs

Note: This table was compiled using publicly available English language information. The "Examples of Ethics-related Activities" were selected because they appeared likely to address a range of different ethical, legal, or social issues, and the category of "Program" includes a variety of organizational structures.

NEUROSCIENCE RESEARCH PROGRAM TITLE	NEUROSCIENCE RESEARCH PROGRAM DESCRIPTION	EXAMPLES OF ETHICS-RELATED ACTIVITIES
Multinational		
Collaborative European Neurotrauma Effectiveness Research in Traumatic Brain Injury (CENTER-TBI)*	• ***Duration***: 2013-2020 • ***Structure***: Collaboration including Belgium, France, Germany, Italy, the Netherlands, Sweden, the United Kingdom, and the United States, among others • ***Purpose(s)***: Conduct traumatic brain injury (TBI) research including clinical research	• "[E]vidence translation and communication" (e.g., improved evidence-based clinical guidelines for the treatment of TBI) • Policy-related activities (e.g., information dissemination to policymakers) • Data sharing policies and procedures
Human Brain Project (HBP)†	• ***Duration***: 2013-2023 • ***Structure***: Collaboration including Argentina, Canada, China, France, Germany, Israel, Japan, the United Kingdom, and the United States, among others • ***Purpose(s)***: Build information and communications technology (ICT) infrastructure for neuroscience, medicine, and computing	• Ethics and Society Programme » Address social, ethical, and philosophical issues » Increase HBP scientists' ethical and social awareness » System of ethical governance
Human Connectome Project (HCP)‡	• ***Duration***: 2010-2015 • ***Structure***: Research consortia, investigators in Germany, Italy, the Netherlands, the United Kingdom, and the United States, among others • ***Purpose(s)***: Conduct structural and functional brain connectivity research	• External Advisory Panel ("scientific and technical advice") • Educational and outreach activities for neuroscientists • Data sharing policies and procedures

* Collaborative European Neurotrauma Effectiveness Research in Traumatic Brain Injury (CENTER-TBI). (2013). Aims [Webpage]. Retrieved April 8, 2014 from https://www.center-tbi.eu/project/aims; CENTER-TBI. (2013). Project Timeline [Webpage]. Retrieved April 8, 2014 from https://www.center-tbi.eu/project/timeline; CENTER-TBI. (2013). Project Participants [Webpage]. Retrieved April 8, 2014 from https://www.center-tbi.eu/project/participants; European Commission. (n.d.) CENTER-TBI. Retrieved from http://ec.europa.eu/research/health/medical-research/brain-research/projects/center-tbi_en.html.

Swiss Broadcasting Corporation. (2013). Brain Project Gets Green Light for EU Funding. Retrieved April 8, 2014 from http:// www.swissinfo.ch/eng/science_technology /Brain_project_gets_ green_light_for_EU _funding.html?cid=34845884; Human Brain Project (HBP). (n.d.). FAQ:

Administration [Webpage]. Retrieved April 8, 2014 from https://www. humanbrainproject.eu /faq/administration; HBP. (2013). Appendix 1: Overall Vision for the Human Brain Project. Retrieved April 8, 2014 from https://www.humanbrainproject.eu/documents/10180 /17646/Vision+Document/8bb75845-8b1d41e0-bcb9-d4de69eb6603; HBP. (2013). The Human Brain Project SP12: Ethics and Society [Video file]. Retrieved April 7, 2014 from http://www.youtube.com/watch?v=0hIbt49VQCs; HBP. (2013). Administration. Retrieved April 7, 2014 from https://www.humanbrainproject.eu/faq/administration.

‡ National Institutes of Health (NIH). (2013). Overview of the Human Connectome Project. Retrieved April 7, 2014 from http:// www.humanconnectome.org/about/project/; NIH. (2010). $40 Million awarded to trace human brain's connections. Retrived from http://www.nih.gov/news/health /sep2010/nimh-15.htm; NIH. (2009). The Human Connectome Project (U54) (RFAMH-10020). Retrieved April 7, 2014 from http://grants.nih.gov/grants/guide/rfa-files/RFA-MH-10-020.html; The NIH Human Connectome Project. (n.d.). Collaborators. Retrieved April 7, 2014 from http://www.humanconnectomeproject.org/about/ collaborators/; Van Essen Lab. (2013). Van Essen Lab Wiki Home Page. Retrieved April 7, 2014 from http://brainvis.wustl.edu/ wiki/index.php/Main_Page.

NEUROSCIENCE RESEARCH PROGRAM TITLE	NEUROSCIENCE RESEARCH PROGRAM DESCRIPTION	EXAMPLES OF ETHICS-RELATED ACTIVITIES
Consortium of Neuroimagers for the Noninvasive Exploration of Brain Connectivity and Tractography (Brain CONNECT)*	• ***Duration***: 2009-2012 • ***Structure***: Consortium of experts on diffusion magnetic resonance imaging (MRI) from Belgium, Denmark, France, Germany, Israel, Italy, Switzerland, the United Kingdom, and the United States • ***Purpose(s)***: Conduct brain morphology and connectivity research	• External Advisory Board • Planned report on awareness and social implications of research
International Neuroinformatics Coordinating Facility (INCF)†	• ***Duration***: 2005-present (as of 2014) • ***Structure***: Consortium of countries including Australia, India, Republic of Korea, Norway, the United Kingdom, and the United States, among others • ***Purpose(s)***: "Develops and maintains [neuroscience] database and computational infrastructure"	• Educational activities aimed predominantly at research community ("to facilitate the flow of information between researchers in academia and industry") • Data sharing policies and procedures
North American Alzheimer's Disease Neuroimaging Initiative (ADNI)‡	• ***Duration***: 2004-2017 • ***Structure***: Research centers in the United States and Canada • ***Purpose(s)***: Conduct early detection and biomarker research, support use of new diagnostics for early detection, strengthen data sharing model	• Data sharing policies and procedures • "ADNI Governance" (ensures compliance with study protocol)

* European Union Seventh Framework Programme. (n.d.) The Connect Consortium. Retrieved April 7, 2014 from http:// cordis.europa.eu/fp7/home_en.html; European Union Seventh Framework Programme. (n.d.). Our Mission [Webpage]. Retrieved April 8, 2014 from http://www.brain-connect.eu/mission.htm; European Union Seventh Framework Programme. (n.d.). News [Webpage]. Retrieved April 7, 2014 from http://www.brain-connect.eu/news.htm; European Union Seventh Framework Programme. (n.d.). MRI and Structural Connectivity. Retrieved April 7, 2014 from http:// www.brain-connect.eu/intro.htm.

† Belgian Federal Science Policy Office (BELSPO). (n.d.). Research Infrastructure. Retrieved April 7, 2014 from https:// www.belspo.be/belspo/coordination/euCoor_Infra_en.stm#INCF; International

Neuroinformatics Coordinating Facility (INCF). (2012). Who We Are [Webpage]. Retrieved April 7, 2014 from http://www.incf.org/about; INCF. (2012). Our Mission [Webpage]. Retrieved April 7, 2014 from http://www.incf.org/about/what-we-do/our-mission; INCF. (2012). INCF National Nodes [Webpage]. Retrieved April 7, 2014 from http://www.incf.org/about/who-we-are/nodes; INCF. (2012). Who We Are [Webpage]. Retrieved April 7, 2014 from http://www.incf.org/about /who-we-are; INCF. (2012). Who Funds Us? [Webpage]. Retrieved April 7, 2014 from http://www.incf.org/about/support-incf/who-funds-us; INCF. (2009). Swedish Foundation for Strategic Research (SSF) Continues INCF Funding. Retrieved April 7, 2014 from http://www.incf.org/newsroom/pressmaterial/news/aggregator/archive/?index=effective& portal_types:list=News%20Item&archive_year:int=2009.

‡ Alzheimer's Disease Neuroimaging Initiative (ADNI). (2013). ADNI Overview [Webpage]. Retrieved April 7, 2014 from http:// www.adni-info.org/Scientists/ADNIOverview.aspx; Alzheimer's Disease Cooperative Study. (n.d.). ADNI 2 Study [Webpage]. Retrieved April 8, 2014 from http://adcs.org/studies/ImagineADNI2.aspx; Alzheimer's Association: Research Center. (n.d.). World Wide Alzheimer's Disease Neuroimaging Initiative [Webpage]. Retrieved April 8, 2014 from http://www.alz.org/ research/funding/partnerships/WW-ADNI_overview.asp. There are numerous regional and country-based ADNIs. See, for example, Europe (http://www.alz.org/research/funding/partnerships/ww-adni_europe.asp); Argentina (http://www.alz.org/research/funding/partnerships/ww-adni_argentina.asp); and Japan (http://www.j-adni.org/etop.html).

NEUROSCIENCE RESEARCH PROGRAM TITLE	NEUROSCIENCE RESEARCH PROGRAM DESCRIPTION	EXAMPLES OF ETHICS-RELATED ACTIVITIES
European Brain Council (EBC)*	• ***Duration***: 2002-present (as of 2014) • ***Structure***: Council includes European neurology, neurosurgery, psychiatry, and neuroscience organizations; patient organizations; and pharmaceutical and biotechnology companies • ***Purpose(s)***: Conduct research on diagnostics and treatments; advocate for increased investment for brain disease research, education, and care; train researchers	• Activities to promote societal dialogue on "neuro-enhancement" (Neuro-Enhancement: Responsible Research and Innovation (NERRI)) • Policy related activities (e.g., information provision to European Commission)
Country-specific (Non-U.S.)		
Norwegian Brain Initiative: A Large-scale Infrastructure for 21st century Neuroscience (NORBRAIN)†	• ***Duration***: 2012-present (as of 2014) • ***Structure***: Norwegian universities • ***Purpose(s)***: Strengthen Norwegian neuroscience research infrastructure	• Educational activities (e.g., researcher education)
Brain Canada‡	• ***Duration***: 2010-present (as of 2014) (successor to the Neuroscience Canada Foundation, est. 1998) • ***Structure***: Board of Directors, Science Advisory Council, and International Science Advisory Council • ***Purpose(s)***: Facilitate research on brain disorders in Canada	• Board of Directors Committees (e.g., "Nominating and Ethics" and "Public Policy and Communications" committees) • Public education about brain disorders in Canada

* European Brain Council (EBC). (n.d.). About Us [Webpage]. Retrieved April 7, 2014 from http://www.europeanbraincouncil. org/about-us/; EBC. (n.d.). Aims [Webpage]. Retrieved April 7,

2014 from http://www.europeanbraincouncil.org/aboutus/aims.asp; EBC. (2006). EBC Funding [Webpage]. Retrieved April 7, 2014 from http://europeanbraincouncil.org/pdfs/ Publications_/EBC%20Funding%20Statement.pdf; EBC. (n.d.). Year of the Brain in Europe-2014. Retrieved April 7, 2014 from http://www.europeanbraincouncil.org/projects/eyob/indeX.asp.

† Norges forskningsråd. (n.d.). Infrastructure for solving challenges in neuroscience [Webpage]. Retrieved April 7, 2014 from http://www.forskningsradet.no/prognett-infrastruktur/Artikkel /Infrastructure_for_solving_challenges_in_ neuroscience/1253962036478.

‡ Brain Canada. (n.d.). Our Mission [Webpage]. Retrieved April 7, 2014 from http://braincanada.ca/en/Our_mission; Brain Canada. (n.d.). Our History [Webpage]. Retrieved April 7, 2014 from http://braincanada.ca/en/Our_history; Brain Canada. (n.d.). Brain Facts [Webpage]. Retrieved April 7, 2014 from http://braincanada.ca/files/NeuroScience_Canada_ Brain_Facts.pdf; Brain Canada. (2011). Annual Report. Retrieved April 7, 2014 from http://braincanada.ca/files/ brainCanadaAnnualReport2011_en.pdf; Brain Canada. (2009). Annual Report 2009. Retrieved April 7, 2014 from http:// braincanada.ca/files/2009AnnualReport.pdf.

APPENDIX II: SUMMARY OF BIOETHICS COMMISSION RECOMMENDATIONS

This first report by the Presidential Commission for the Study of Bioethical Issues (Bioethics Commission), as part of the Brain Research through Advancing Innovative Neurotechnologies (BRAIN) Initiative, emphasizes the importance of integrating ethics into neuroscience research from the earliest planning phases. In a second report, the Bioethics Commission will consider the ethical and societal implications of neuroscience research and its applications more broadly—ethical implications that a strongly integrated research and ethics infrastructure will be well equipped to address, and that myriad stakeholders, including scientists, ethicists, educators, public and private funders, advocacy organizations, and the public must be prepared to handle. In this report, the Bioethics Commission concludes that ethics and neuroscience research should be integrated throughout the research endeavor and offers four recommendations to facilitate the successful integration of ethics and neuroscience research.

Ethics integration should aid scientists in recognizing and addressing ethical issues as they arise, and ethicists in understanding the science and technology with which they engage. Although ethics is already integrated into science in various ways—as it is in many human activities—more explicit and systematic integration serves to elucidate implicit ethical judgments and allows their merits to be assessed thoughtfully.

Recommendation 1:

Integrate Ethics Early and Explicitly Throughout Research

Institutions and individuals engaged in neuroscience research should integrate ethics across the life of a research endeavor, identifying the key ethical questions associated with their research and taking immediate steps to make explicit their systems for addressing those questions. Sufficient resources should be dedicated to support ethics integration.

Once implemented, systems for integrating ethics and neuroscience research should not operate in a vacuum. Institutions and individuals engaged in neuroscience research should learn from collective experience to improve existing systems and inform others about what works and what does not.

Recommendation 2:
Evaluate Existing and Innovative Approaches to Ethics Integration
Government agencies and other research funders should initiate and support research that evaluates existing as well as innovative approaches to ethics integration. Institutions and individuals engaged in neuroscience research should take into account the best available evidence for what works when implementing, modifying, or improving systems for ethics integration.

One foundational approach to integration is pairing science and ethics education at all levels of education. Early ethics education in academic settings is critical to prepare future scientists to integrate ethical considerations into their work—including future research in neuroscience. Professional development for experienced investigators is equally important and can serve multiple ends, contributing not only to their individual knowledge, but to the knowledge of the students and young scientists that they mentor as well.

Recommendation 3:
Integrate Ethics and Science through Education at All Levels
Government agencies and other research funders should initiate and support research that develops innovative models and evaluates existing and new models for integrating ethics and science through education at all levels.

It has been just over a year since the announcement of the BRAIN Initiative, and institutions participating in this research effort have an important opportunity to integrate ethics and science from the outset. A key component of this integration is the inclusion of ethicists or scientists with experience in ethics in BRAIN Initiative-related scientific advisory boards and funding review committees, particularly for the major public and private sector partners.

Recommendation 4:
Explicitly Include Ethical Perspectives on Advisory and Review Bodies
BRAIN Initiative-related scientific advisory and funding review bodies should include substantive participation by persons with relevant expertise in the ethical and societal implications of the neuroscience research under consideration.

APPENDIX III: GUEST PRESENTERS TO THE BIOETHICS COMMISSION REGARDING ETHICS AND NEUROSCIENCE

William D. Casebeer, Ph.D.
(U.S.A.F., Retired)
Program Manager, Defense Advanced Research Projects Agency
Timothy Caulfield, LL.M., F.R.S.C., F.C.A.H.S.
Canada Research Chair in Health Law and Policy; Professor in the Faculty of Law and the School of Public Health, University of Alberta

David Chalmers, Ph.D.
Professor of Philosophy and Co-director of the Center for Mind, Brain, and Consciousness, New York University; Distinguished Professor of Philosophy and Director of the Centre for Consciousness, Australian National University

Anjan Chatterjee, M.D., F.A.A.N. Professor of Neurology, Center for Cognitive Neuroscience and Center for Functional Neuroimaging, University of Pennsylvania School of Medicine

Mildred Cho, Ph.D.
Associate Director; Professor of Pediatrics, Stanford Center for Biomedical Ethics, Stanford University

Miyoung Chun, Ph.D.
Executive Vice President of Science Programs, The Kavli Foundation

Martha Farah, Ph.D.
Walter H. Annenberg Professor in Natural Sciences; Professor of Psychology; Director, Center for Cognitive Neuroscience; Director, Center for Neuroscience and Society;
Senior Fellow, Center for Bioethics University of Pennsylvania

Erik Fisher, Ph.D.
Associate Director for Integration, Center for Nanotechnology in Society; Assistant Professor, School of Politics and Global Studies and the Consortium for Science, Policy and Outcomes, Arizona State University

Hank Greely, J.D.
Deane F. and Kate Edelman Johnson Professor of Law, Stanford Law School; Professor (by courtesy) of Genetics, Stanford Medical School; Chair, Steering Committee of the Center for Biomedical Ethics; Director, Center for Law and the Biosciences; Director, Stanford Interdisciplinary Group on Neuroscience and Society and its Program in Neuroethics, Stanford Law School

Barbara Herr Harthorn, Ph.D.
Director, NSF Center for Nanotechnology in Society; Professor, Department of Anthropology, University of California, Santa Barbara

Steven E. Hyman, M.D.
Founding President, International Neuroethics Society; Director, Stanley Center for Psychiatric Research, Broad Institute of Massachusetts Institute of Technology and Harvard University

Judy Illes, Ph.D.
Professor of Neurology;
Canada Research Chair in Neuroethics; Director, National Core for Neuroethics; Faculty, Brain Research Centre, University of British Columbia

Deborah G. Johnson, Ph.D., M.Phil., M.A.
Anne Shirley Carter Olsson Professor of Applied Ethics, Science, Technology, and Society Program, Department of Engineering and Society, School of Engineering and Applied Science, University of Virginia

Christof Koch, Ph.D. Chief Scientific Officer,
Allen Institute for Brain Science

Walter J. Koroshetz, M.D.
Deputy Director, National Institute of Neurological Disorders and Stroke, National Institutes of Health

Pat Levitt, Ph.D.
Chair-Elect, Neuroscience Section, American Association for the Advancement of Science; Provost Professor, Department of Pediatrics; W.M. Keck Chair in Neurogenetics, Keck School of Medicine, University of Southern California; Director, Program in Developmental Neurogenetics, Institute for the Developing Mind, Children's Hospital Los Angeles

Bernard Lo, M.D.
Professor of Medicine;
Director, Program in Medical Ethics, University of California, San Francisco

Peggy Mason, Ph.D.
Chair, Ethics Committee, Society for Neuroscience; Professor, Department of Neurobiology, University of Chicago

Jonathan Montgomery, LL.M. Chair, Nuffield Council on Bioethics; Professor of Health Care Law, University College London

Stephen Morse, J.D., Ph.D.
Ferdinand Wakeman Hubbell Professor of Law; Professor of Psychology and Law in Psychiatry, University of Pennsylvania Law School and School of Medicine

Thomas H. Murray, Ph.D.
President Emeritus, The Hastings Center

Ushma Neill, Ph.D.
Director, Office of the President, Memorial Sloan-Kettering Cancer Center; Editor at Large, Journal of Clinical Investigation

John Perry, Ph.D.
Emeritus Professor of Philosophy, Stanford University; Distinguished Professor of Philosophy,
University of California, Riverside

Eric Racine, Ph.D.
Director, Neuroethics Research Unit; Associate Research Professor, Institut de Recherches Cliniques de Montréal; Associate Research Professor, Department of Medicine, Université de Montréal; Adjunct Professor, Department of Medicine and Department of Neurology and Neurosurgery, McGill University

Nikolas Rose, Ph.D.
Member, Human Brain Project Social and Ethical Division Steering Committee; Professor of Sociology and Head of Department of Social Science, Health and Medicine, King's College London

Bruce R. Rosen, M.D., Ph.D. Professor of Radiology, Harvard Medical School; Director, Athinoula A. Martinos Center for Biomedical Imaging, Department of Radiology, Massachusetts General Hospital

Adina Roskies, Ph.D.
Associate Professor, Department of Philosophy, Dartmouth College

Pamela Sankar, Ph.D.
Associate Professor, Department of Medical Ethics and Health Policy; Senior Fellow, Leonard Davis Institute of Health Economics, University of Pennsylvania

Marya Schechtman, Ph.D. Professor of Philosophy, University of Illinois at Chicago

Terrence J. Sejnowski, Ph.D. Francis Crick Chair; Professor and Laboratory Head, Computational Neurobiology Laboratory, Salk Institute for Biological Studies; Investigator, Howard Hughes Medical Institute; Distinguished Professor, Section of Neurobiology/Neurosciences, University of California, San Diego

Stefano Semplici, Ph.D.

Chairperson, International Bioethics Committee, United Nations Educational, Scientific, and Cultural Organization; Professor of Social Ethics, University of Rome Tor Vergata

Mildred Z. Solomon, Ed.D.

President and CEO, The Hastings Center; Clinical Professor of Anaesthesia, Harvard Medical School

Nicholas Steneck, Ph.D.

Director, Research Ethics and Integrity Program, Michigan Institute for Clinical and Health Research; Professor Emeritus of History, University of Michigan

Anthony Wagner, Ph.D.

Professor of Psychology and Neuroscience, Stanford University

Stephen J.A. Ward, Ph.D.

Professor; Director, George S.

Turnbull Center, School of Journalism and Communication, University of Oregon-Portland

John C. Wingfield, Ph.D.

Assistant Director for the Directorate for Biological Sciences, National Science Foundation

Paul Root Wolpe, Ph.D.

Director, Center for Ethics; Asa Griggs Candler Professor of Bioethics, Emory University

David E. Wright, Ph.D.

Director, Office of Research Integrity, U.S. Department of Health and Human Services

About the Presidential Commission for the Study of Bioethical Issues

The Presidential Commission for the Study of Bioethical Issues (Bioethics Commission) is an advisory panel of the nation's leaders in medicine, science, ethics, religion, law, and engineering. The Bioethics Commission advises the President on bioethical issues arising from advances in biomedicine and related areas of science and technology. The Bioethics Commission seeks to identify and promote policies and practices that ensure scientific research, health care delivery, and technological innovation are conducted in a socially and ethically responsible manner.

RESIDENTIAL COMMISSION FOR THE STUDY OF BIOETHICAL ISSUES

AMY GUTMANN, Ph.D., CHAIR
President and Christopher H. Browne
Distinguished Professor of Political Science and Professor of Communication, University of Pennsylvania

JAMES W. WAGNER, Ph.D., VICE CHAIR
President, Emory University

ANITA L. ALLEN, J.D., Ph.D.
Vice Provost for Faculty, Henry R. Silverman Professor of Law and Professor of Philosophy, University of Pennsylvania

JOHN D. ARRAS, Ph.D.
Porterfield Professor of Biomedical Ethics, Professor of Philosophy, Professor of Public Health Sciences, University of Virginia

BARBARA F. ATKINSON, M.D.
Executive Vice Chancellor Emeritus, University of Kansas Medical Center; Professor Emeritus of Pathology and Laboratory Medicine, University of Kansas School of Medicine

NITA A. FARAHANY, J.D., Ph.D.
Director of Science and Society, Professor of Law and Philosophy, Professor of Genome Sciences and Policy, Duke University

CHRISTINE GRADY, R.N., Ph.D.
Chief, Department of Bioethics, National Institutes of Health Clinical Center

STEPHEN L. HAUSER, M.D.
Robert A. Fishman Distinguished Professor and Chair of the Department of Neurology, University of California, San Francisco

RAJU S. KUCHERLAPATI, Ph.D.
Paul C. Cabot Professor, Department of Genetics, Harvard Medical School; Professor, Department of Medicine, Brigham and Women's Hospital

NELSON L. MICHAEL, M.D., Ph.D.
Colonel, Medical Corps, U.S. Army; Director, U.S. Military HIV Research Program, Walter Reed Army Institute of Research

DANIEL P. SULMASY, M.D., Ph.D., FACP
Kilbride-Clinton Professor of Medicine and Ethics, Department of Medicine and Divinity School; Associate Director, The MacLean Center for Clinical Medical Ethics, University of Chicago

PRESIDENTIAL COMMISSION FOR THE STUDY OF BIOETHICAL ISSUES
STAFF* AND CONSULTANTS

Executive Director
Lisa M. Lee, Ph.D., M.S.

Associate Directors
Michelle Groman, J.D.
Kayte Spector-Bagdady, J.D., M. Bioethics

Communications Director
Hillary Wicai Viers, M.S.J.

Senior Advisors
Paul A. Lombardo, Ph.D., J.D.
Jonathan D. Moreno, Ph.D.

Research Staff

Misti Ault Anderson, M.S., M.A.
Rachel S. Bressler, J.D.
Kata Chillag, Ph.D.
Elizabeth M. Fenton, Ph.D., M.P.H.
Karen M. Meagher, Ph.D.
Olivia Nevitt, M.P.H.
Cristina Nigro, M.S.
Elizabeth R. Pike, J.D., LL.M.
Maneesha Sakhuja, M.H.S.
Michelle Spektor, B.S.
Nicolle K. Strand, J.D., M. Bioethics
Michael N. Tennison, M.A.
Victoria Wilbur, B.A.
Tenny R. Zhang, B.A.

Consultants
Burness Communications
Kathi E. Hanna, M.S., Ph.D.

Administrative Staff
Tynetta Dreher
Esther E. Yoo

Interns
Ruqayyah Abdul-Karim, B.A.
Cassandra Hunter, M.A.
Rahul Nayak, B.S.

Abbas Rattani, M. Bioethics
Kevin Tobia, B.A.
Sean Valle, B.A.
Abena Yeboa, B.S., B.A.

**Includes former research staff*

In: Ethical Issues in Neuroscience Research
Editor: Marilyn Phillips
ISBN: 978-1-63482-989-2

Chapter 2

Gray Matters: Topics at the Intersection of Neuroscience, Ethics, and Society*

Presidential Commission for the Study of Bioethical Issues

Executive Summary

Neuroscience presents an unparalleled opportunity to gain a deeper understanding of the human brain and mind, including our cognition, behavior, memory, learning, mood, and social interactions. It also offers new opportunities to treat, prevent, and possibly cure neurological disorders that constitute an immense public health burden worldwide.

In 2013, President Obama announced the federal Brain Research through Advancing Innovative Neurotechnologies (BRAIN) Initiative, and charged the Presidential Commission for the Study of Bioethical Issues (Bioethics Commission) to examine ethical considerations related to both the conduct of neuroscience research and the application of neuroscience research findings. The Bioethics Commission addressed the President's charge in two parts. In its first volume on neuroscience and ethics, *Gray Matters: Integrative Approaches for Neuroscience, Ethics, and Society*, the Bioethics Commission emphasized the importance of integrating ethics and neuroscience throughout the research endeavor.[1] This second volume, *Gray Matters: Topics at the Intersection of Neuroscience, Ethics, and Society*, takes an in-depth look at three topics at the intersection of neuroscience and society that have captured the public's attention.

The Bioethics Commission found widespread agreement that contemporary neuroscience holds great promise for relieving human suffering from a number of devastating neurological disorders. Less agreement exists on multiple other topics, and the Bioethics Commission focused on three cauldrons of controversy—cognitive enhancement, consent capacity, and neuroscience and the legal system. These topics illustrate the ethical tensions and societal implications of advancing neuroscience and technology, and bring into heightened relief many important ethical considerations.

* This is an edited, reformatted and augmented version of Volume 2 of a two volume report, issued March 2015.

The Bioethics Commission seeks to clarify for the public the scientific landscape, identify common ground for productive discourse surrounding these topics, and recommend an ethical path forward to support the progress of neuroscience research.

Cognitive Enhancement and Beyond

In this report, the Bioethics Commission expanded the conversation beyond the ongoing cognitive enhancement debate to include other forms of neural modification. First, it considered not only novel neurotechnologies, but also methods, behaviors, and conditions that alter the brain and nervous system. Second, it expanded the discussion beyond use of products and methods that enhance cognition to include those that alter the brain or nervous system in a wide range of ways, such as altering motor function.

Neural modification can serve at least three purposes, to (1) *maintain or improve* neural health and cognitive function within typical or statistically normal ranges; (2) *treat* disease, deficiency, injury, impairment, or disorder (referred to as "neurological disorders") to achieve or restore typical or statistically normal functioning; and (3) *expand or augment* function above typical or statistically normal ranges. In delineating these neural modification objectives, the Bioethics Commission is mindful that they are not always sharply distinguishable.

Altering the brain and nervous system is not inherently ethical or unethical. Ethical assessment of neural modification requires consideration of who is choosing the modifier, what is being chosen, what its purposes are, who stands to benefit, and who might be harmed. Members of the public must be well-informed to make educated, practical decisions about personal health and wellbeing, and participate in collective deliberation and decision making about societal applications of neural modifiers.

Several well-known lifestyle interventions, such as adequate sleep, exercise, and nutrition, are associated with improved neural function. Similarly, public health interventions, such as lead paint abatement, can help prevent the negative impact of environmental exposures on neural development and function.[2] These behaviors and conditions can maintain and improve neural health and might be safer and more effective than those offered by novel neural modifiers, some of which might have minimal benefits and uncertain risks.[3]

Recommendation 1: Prioritize Existing Strategies to Maintain and Improve Neural Health

In addition to developing new drugs and devices to maintain and improve neural health, funders should prioritize and support research on existing, low-technology strategies, such as healthy diet, adequate exercise and sleep, lead paint abatement, high-quality educational opportunities, and toxin-free workplaces and housing.

Existing treatments for neurological disorders are valuable and can be improved. Emerging neural modification interventions will help reduce the individual and societal burden of neurological disorders. Safe and effective treatments can improve the lives of millions of individuals living with such conditions.

Recommendation 2: Prioritize Treatment of Neurological Disorders

Funders should prioritize research to treat neurological disorders to improve health and alleviate suffering. This research should consider individual, familial, and public health burdens as well as potential risks, benefits, and long-term effects of specific interventions.

Although the Bioethics Commission recognizes the need to prioritize the study of both traditional and novel interventions for the prevention and treatment of neurologic disorders, it nonetheless also supports research to better characterize and understand novel neural modification techniques to augment or enhance neural function. Limited, inconclusive evidence exists for the benefits and risks of stimulant drugs and brain stimulation methods as neural enhancers.[4] In addition, few data are available on the prevalence of the use of neural modification interventions for cognitive enhancement purposes.

Recommendation 3: Study Novel Neural Modifiers to Augment or Enhance Neural Function

Funders should support research on the prevalence, benefits, and risks of novel neural modifiers to guide the ethical use of interventions to augment or enhance neural function.

If safe and effective novel forms of cognitive enhancement become available, they will present an opportunity to insist on a distribution that is fair and just. While not eliminating all other less tractable forms of injustice in the distribution of neural health and wellbeing, it is possible to ensure that any new forms of safe and beneficial neural modification do not worsen those injustices.

Limiting access to effective enhancement interventions to those who already enjoy greater access to other social goods would be unjust. It also might deprive society of other benefits of more widespread enhancement that increase as more individuals have access to the intervention.[5] In addition, more widespread enhancement might help to close some gaps in opportunity that are related to neural function, such as educational attainment or employment.

Recommendation 4: Ensure Equitable Access to Novel Neural Modifiers to Augment or Enhance Neural Function

Policymakers and other stakeholders should ensure that access to beneficial, safe, effective, and morally acceptable novel neural modifiers to augment or enhance neural function is equitable so as not to compound or exacerbate social and economic inequities.

Clinicians often receive requests to prescribe medications for cognitive enhancement, and they must decide whether to prescribe the medication to particular patients.[6] These decisions are more ethically complex with regard to children, because children lack legal and ethical consent capacity and are vulnerable to coercion.[7] Other stakeholders also would benefit from education and guidance on neural modification interventions. These stakeholders include employers, parents, educators, and professional organizations in fields such as aviation, medicine, and the military, among others, that are associated with on-the-job use of brain and nervous system enhancement interventions.

Recommendation 5: Create Guidance About the Use of Neural Modifiers

Professional organizations and other expert groups should develop guidance for clinicians, employers, parents, educators, and patients about the use of neural modifiers and their potential risks and benefits. Medical professional organizations should develop guidelines to assist clinicians in responding to requests for prescriptions for interventions to expand or augment neural function. Clinicians should not prescribe medications that have uncertain or unproven benefits and risks to augment neural function in children and adolescents who do not have neurological disorders.

Capacity and the Consent Process

Neuroscientists who conduct research involving human participants commonly work with populations or individuals whose consent capacity might be absent, impaired, fluctuating, or in question.[8] Understanding, evaluating, and improving informed consent processes is an ongoing goal of research ethics. The history of research ethics includes multiple efforts by national-level advisory bodies to provide guidance for research involving individuals who might have compromised or impaired consent capacity. This history illustrates the challenging tension between the need for rigorous research on important diseases and conditions, and the need to protect individuals who might be vulnerable because of impaired consent capacity. Federal regulations specific to research involving individuals with impaired consent capacity have never been adopted.

Neuroscience research is an important means of promoting progress and benefiting populations affected by neurological disorders and psychiatric conditions, including those associated with impaired consent capacity, and should proceed with adequate ethical safeguards and protections in place. This dual mission—protection and inclusion to ensure that the benefits of research are distributed equitably—shapes many core ethical considerations surrounding capacity, the consent process, and participation in research.

Recommendation 6: Responsibly Include Participants with Impaired Consent Capacity in Neuroscience Research

Researchers should responsibly include individuals with impaired consent capacity who stand to benefit from neuroscience research. Participation, with ethical safeguards in place, can ensure progress aimed at understanding and ameliorating neurological disorders and psychiatric conditions.

Researchers have made substantial progress in the past decade in characterizing and understanding consent capacity. However, gaps remain, and further research can support development of best practices for ethical research involving participants with impaired consent capacity. Conceptual research on gaps in our knowledge, including the influence of vulnerability, desperation, and affective states on decision making, could lead to better protections for all research participants. Moreover, empirical research evaluating assessment tools and additional protections for participants with impaired consent capacity can determine whether they are adequately protective.

Recommendation 7: Support Research on Consent Capacity and Ethical Protections

Funders should support research to address knowledge gaps about impaired consent capacity, including the concept of capacity, brain function and decision-making capacity, current policies and practices, and assessment tools.

Equating certain conditions with impaired consent capacity or making unfounded assumptions about individual abilities based on diagnoses can exacerbate or perpetuate stigma. Ethical neuroscience research can foster a more accurate understanding of neurological disorders and mental illness, and potentially mitigate stigma. One principal approach to help neuroscience researchers alleviate stigma is stakeholder engagement.

Recommendation 8: Engage Stakeholders to Address Stigma Associated with Impaired Consent Capacity

Funders and researchers should engage stakeholders, including members of affected communities, to build understanding of consent capacity and associated diagnoses to mitigate the potential for stigma and discrimination.

Including affected individuals (those with impaired consent capacity and others) in research is vital to fulfill the promise of neuroscience to ameliorate neurological disorders and psychiatric conditions. The Common Rule—the regulations supplying standards for the ethical conduct of federally supported human subjects research—requires informed consent from research participants or legally authorized representatives (LARs) before research can proceed.[9] Thus, an important step in conducting ethical research involving individuals with impaired consent capacity is determining who can serve as an LAR. Federal and state laws lack clarity about how to make this determination.

Recommendation 9: Establish Clear Requirements for Identifying Legally Authorized Representatives for Research Participation

State legislatures and federal regulatory bodies should establish clear requirements to identify who can serve as legally authorized representatives for individuals with impaired consent capacity to support their responsible inclusion in research.

Neuroscience and the Legal System

Advances in neuroscience offer a better understanding of human behavior, the potential for improved policymaking, increased accuracy, and decreased errors in advancing justice. The application of neuroscience to the law also raises concerns about scientific reliability, misapplication and overreliance on a developing science, conceptions of free will, mental privacy, and personal liberty. Although neuroscience might help us achieve more accuracy in decision making and better policies for trials and sentencing, it does not change the normative or moral questions that the law seeks to answer. Law is a social institution, built on norms developed and instituted by society.

Neuroscience has multiple potential applications to the legal system and already is employed in many relevant contexts. Prosecutors and defense attorneys use neuroscience evidence in criminal proceedings to support propositions concerning, for example, competency to stand trial, mitigation of criminal responsibility, and predicting future dangerousness. Parties also use neuroscience evidence in the civil context to provide objective evidence of "invisible" injuries such as toxic exposure, pain, and suffering. Policymakers have invoked neuroscience to advocate for legislation and reform; scholars have advocated for use of neuroscience to address biases in legal decision making; and researchers and even some commercial entities have introduced novel uses of neuroscience for investigative purposes.

Members of the public, especially ones who will serve as jurors, can benefit immensely from educational resources that help bring high-level neuroscientific concepts into lay terms. Individuals expected to use and interpret neuroscience, including judges and attorneys, can also benefit from greater availability of basic training that helps ease the interdisciplinary transition of neuroscience into the legal decision-making process and effectively assess the evidence and technologies involved in a growing number of legal cases.

Recommendation 10: Expand and Promote Educational Tools to Aid Understanding and Use of Neuroscience within the Legal System

Government bodies and professional organizations, including legal societies and nonprofit organizations, should develop, expand, and promote training resources, primers, and other educational tools that explain the application of neuroscience to the legal system for distribution to members of the public, jurors, judges, attorneys, and others.

In addition to the broad educational tools discussed in Recommendation 10, relevant bodies also should fund and conduct specific research and report results regarding use of neuroscience evidence in making important legal and policy decisions. Organizations and government bodies also should publish reports that address the challenges and limitations of neuroscience's application to the legal system.

Recommendation 11: Fund Research on the Intersection of Neuroscience and the Legal System

Relevant bodies, such as the National Academies of Science, the U.S. Department of Justice, the National Institute of Justice, and the Social Security Administration, should support comprehensive studies of the use of neuroscience in legal decision making and policy development.

Neuroscience can add value to legal decision making and policy development. To maximize the value that neuroscience has to offer, scientists, the media, and legal decision makers must avoid hype. When neuroscience evidence that is unreliable or has not yet been validated and is not ready for application is introduced into the legal system, justice is threatened. Unrealistically high expectations for new science and technology can lead to a loss of trust when those expectations are unmet.[10]

Recommendation 12: Avoid Hype, Overstatement, and Unfounded Conclusions

Neuroscientists, attorneys, judges, and members of the media should not overstate or rely too heavily on equivocal neuroscientific evidence to draw conclusions about behavior, motivations, intentions, or legal inferences.

As attorneys introduce more neuroscience evidence into the courtroom, and advocates use neuroscience to influence policy, neuroscientists should engage with the process, consider potential legal applications of their work, and seek to engage with legal and policy decision makers to ease the translation. Neuroscientists can play a principal role in assisting judges and jurors with determining the appropriate interpretations of neuroscientific evidence.[11]

Recommendation 13: Participate in Legal Decision-Making Processes and Policy Development

Neuroscientists should participate in legal decision-making processes and policy development to ensure the accurate interpretation and communication of neuroscience information.

CONCLUSION

In this report, the Bioethics Commission calls for research on a number of critical topics. Such research requires adequate support, including funding, personnel, and other resources. As a White House Grand Challenge, the BRAIN Initiative is uniquely positioned to establish and support efforts that bring together diverse expertise from neuroscience, ethics, law, policy, and other disciplines to advance research and education at the intersection of neuroscience, ethics, and society.

Recommendation 14: Establish and Fund Multidisciplinary Efforts to Support Neuroscience and Ethics Research and Education

The BRAIN Initiative should establish and fund organized, independent, multidisciplinary efforts to support neuroscience and ethics research and education, including the activities recommended in this report.

* * *

Neuroscience advances have captured the public's attention and stimulated scholarly and public debate, fueled by accurate accounts of the science as well as hyped or misinformed interpretations. Three controversies are some of the most important and provocative topics at the intersection of neuroscience and society. Neural modification, including cognitive enhancement, raises questions about reconciling risks and benefits, ensuring justice, and understanding what it means to be human. Adequately respecting and protecting individuals with impaired consent capacity has presented challenges for decades. Advances in neuroscience and the promise of neuroscience research compel us to reexamine this area—ensuring that those with impaired consent capacity can participate in and benefit from ethical research. Application of neuroscience to legal decision making and policy development offers potential for more accurate and just outcomes, but also raises concerns about premature use of scientific information, privacy, and moral responsibility. In this report, the Bioethics Commission seeks to clarify the scientific landscape, identify common ground, and recommend ethical paths forward to stimulate and continue critical, well-informed conversations at the intersection of neuroscience and ethics as the field continues to advance.

SECTION 1. INTRODUCTION

Advances in contemporary neuroscience research offer the prospect of great individual and societal benefit. Neuroscience presents an unparalleled opportunity to gain a deeper understanding of the human brain and mind, including our cognition, behavior, memory, learning, mood, and social interactions. It also offers new opportunities to treat, prevent, and possibly cure neurological disorders that constitute an immense public health burden worldwide. In 2013, President Obama announced the federal Brain Research through Advancing Innovative Neurotechnologies (BRAIN) Initiative, harnessing a diverse array of experts to pursue discoveries that will have a "lasting positive impact on lives, the economy, and our national security."[12] The President charged the Presidential Commission for the Study of Bioethical Issues (Bioethics Commission) to examine ethical considerations related to both the conduct of neuroscience research and the application of neuroscience research findings.

The Bioethics Commission addressed the President's charge in two parts. In its first volume on neuroscience and ethics, *Gray Matters: Integrative Approaches for Neuroscience, Ethics, and Society* (*Gray Matters*, Vol. 1), the Bioethics Commission emphasized the importance of integrating ethics and neuroscience throughout the research endeavor.[13] This second volume, *Gray Matters: Topics at the Intersection of Neuroscience, Ethics, and Society* (*Gray Matters*, Vol. 2), takes an in-depth look at three particularly controversial topics at the intersection of neuroscience and society that have captured the public's attention.

After engaging experts and reviewing public comments, the Bioethics Commission found widespread agreement that contemporary neuroscience holds great promise for relieving human suffering from numerous devastating neurological disorders. Less agreement exists on multiple other topics, and the Bioethics Commission resolved to focus its second *Gray Matters* volume on three cauldrons of controversy—cognitive enhancement, consent capacity, and neuroscience and the legal system. These three topics illustrate the ethical tensions and societal implications of advancing neuroscience and technology, and bring into heightened relief many important ethical considerations.

These controversial topics sit at the epicenter of both scholarly debate and public dialogue in popular media about the reach and impact of neuroscience.[14] Alongside well-informed and well-intentioned voices, hyperbole and misinformation permeate the conversation—this exaggeration can lead to undue excitement and attention, commonly referred to as "hype." Overstated interpretation of research results and expectations can mislead the public, misdirect resources, and instill misplaced fears. In this report, the Bioethics Commission aims to clear a path for productive discourse and the ethical progress of neuroscience research—progress that is crucial to alleviating serious medical conditions and bettering society in numerous other ways. Drawing from its collective expertise and crucial role as public educator, the Bioethics Commission offers this report to clarify for the public the scientific landscape, identify common ground, and recommend ethical paths forward on these topics.

Cognitive Enhancement and Beyond

In universities and colleges, doctors' offices, secondary schools, professional societies, and popular newspapers and blogs, scholars, professionals, caregivers, and others actively debate the use of novel neuroscience products and other methods to enhance cognition. In this report, the Bioethics Commission clarifies and expands the conversation beyond the common cognitive enhancement debate. First, this report considers not only novel neurotechnologies, but also methods, behaviors, and conditions that alter the brain and nervous system. Second, the discussion moves beyond the use of products and methods that enhance *cognition* to include those that alter the brain or nervous system in wide-ranging ways, such as altering motor function.

The definition of enhancement and the distinction between enhancement and other forms of neural modification are at the center of this debate. For example, *The Neuroethics Blog* hosted by the Center for Ethics in the Neuroethics Program at Emory University (Atlanta, Georgia) regularly posts brief pieces covering topics at the intersection of neuroscience and ethics. A 2014 post outlined the difficulty in defining "normal" and distinguishing treatment and enhancement.[15] Another post questioned whether diminishment of certain capabilities can be considered a form of enhancement.[16]

Journalists also have reported on other ethical aspects of cognitive enhancement. For example, news stories have raised questions about both ethical and scientific controversy surrounding cognitive enhancement, including whether the wakefulness-promoting drug modafinil (Provigil®) boosts cognition, whether transcranial direct current stimulators can or should be regulated by the U.S. Food and Drug Administration (FDA), whether it is ethically permissible to erase bad memories, and whether brain augmentation is ethically justifiable.[17]

Scholars and the public also express concern over the use of direct-to-consumer (DTC) neuroscience products purported to enhance cognition. A 2014 consensus statement issued by a consortium of scientists rejects the claim that currently marketed brain games reduce or reverse cognitive decline, and several media outlets reported on the consensus.[18] In addition, a report on one DTC neuroscience company's announcement that it would allow researchers access to its vast repository of data collected from user play of its brain training games included skepticism about the accuracy of the data and possible inadequacy of privacy protections.[19]

Several stakeholders, including national and international leaders in neuroscience and ethics, urged the Bioethics Commission to consider cognitive enhancement and related topics. For example, the International Neuroethics Society considers human enhancement to be one of the most important ethical considerations related to neuroscience, focusing its comments to the Bioethics Commission on justice concerns, product safety and effectiveness, potential impact of DTC products, and implications for use of neural modifiers with healthy children and adolescents.[20] Against a background of strongly held and often opposing views, cognitive enhancement and neural modification are ripe for the Bioethics Commission to consider.

ETHICS INTEGRATION AND MITIGATING HYPE

Contemporary neuroscience offers great promise for attaining a deeper understanding of the human brain and alleviating the burden of neurological disorders, but overstatement and hype persist. Neuroscientists, members of the media who report on neuroscience advances, and others sometimes make claims that can be false, misleading, distorted, or exaggerated.

For example, the *New York Times Magazine* published an article in January 2015 about a researcher attempting to map all of the neural connections in the brain. The magazine's cover depicts the words "This Is Your Brain" in letters formed by illustrated neurons. Prominent captions on the magazine's cover and article cover page suggest that we could learn "everything" about ourselves through such a map, and ask, "If he succeeds, could we live forever as data?"

Less prominently, in its final paragraphs, the article presents critiques from scientists who argue that brain mapping would be "absolutely necessary but completely insufficient" to fully understand the human brain.

In *Gray Matters*, Vol. 1, the Bioethics Commission emphasized the importance of integrating ethics and neuroscience throughout the research endeavor, in part by integrating the perspectives of philosophers, social scientists, and others into the research process. An integrated perspective on the research described in this article might point out that even a complete map of the human brain would not fully inform our understanding of consciousness and the self. Articles that feature attention-grabbing headlines, but lack complete and balanced perspectives, serve to inform and excite the public about important neuroscience research, but run the risk of distorting and exaggerating its actual potential.

Sources: Cook, G. (2015, January 11). Mind Games. *New York Times Magazine*, pp. 27-31, 50; Presidential Commission for the Study of Bioethical Issues (PCSBI). (2014, May). *Gray Matters: Integrative Approaches for Neuroscience, Ethics, and Society*. Washington, DC: PCSBI.

Capacity and the Consent Process

The burden of neurological disorders, mental illnesses, brain and nervous system injuries, and age-related cognitive decline is enormous. Progressing through human life without being affected by a decline in brain or nervous system function or needing to care for an affected loved one is likely impossible. Neuroscience offers great promise to reduce this burden and potentially find ways to prevent, treat, and cure many brain-related disorders and injuries. Clinical research involving affected populations is necessary to achieve this worthy goal. Managing a robust informed consent process with individuals who might have impaired consent capacity is especially pertinent in contemporary neuroscience. Impaired consent capacity is a hotly debated topic in research ethics, yet regulations, guidance, and research practices have remained essentially unchanged for decades. The Bioethics Commission is poised to move the conversation forward.

Although members of the public might not explicitly consider the topic of consent capacity, they are deeply affected by it. Even the healthy among us might one day benefit from interventions that preserve life or neurological function, borne from the fruits of neuroscience research that requires enrollment of individuals with impaired consent capacity. The popular media reported on some of the scholarly debate around this topic. For example, media sources reported on the FDA's informed consent draft guidance, released in 2014, acknowledging that it addresses individuals "with diminished or fluctuating consent capacity."[21] Others criticized the FDA's position on legally authorized representatives in light of a dearth of state laws indicating who can legally serve as a representative in research.[22] A 2014 post on *The Neuroethics Blog* considers the elements of ethical informed consent from individuals who might have impaired consent capacity in situations when the investigator stands to gain financially but the research participant will not benefit from the research.[23]

Several stakeholders urged the Bioethics Commission to weigh in on the consent capacity discussion. For example, Public Responsibility in Medicine and Research (PRIM&R) identified numerous reports and recommendations from national bodies on consent capacity, inquired as to why many of those recommendations have not been implemented, and urged consistency in the field.[24] Given the ongoing challenge to respect and adequately protect those with impaired consent capacity—who stand to gain the most, but who also would shoulder much of the burden of research—and the potential for contemporary neuroscience to help guide ethical policies, the Bioethics Commission addresses the topic in this report.

Neuroscience and the Legal System

The fruits of neuroscience research are applied in a variety of contexts, increasingly extending beyond the laboratory, from clinical translation to development of DTC products. Use of neuroscience in legal decision making and policy development is a controversial matter that has captured the public's attention. Application of neuroscience to the legal system ranges from using evidence about the development of adolescent brains to prevent a teenager from receiving the death penalty, to using imaging technology to persuade a jury that a defendant might not be capable of the moral responsibility required for conviction of a crime.

Use of neuroscience within the legal system raises fundamental questions about how to determine when scientific findings are ready for public use, and how to ensure that scientific experts who testify and play a role in determining the fate of defendants are reliable and their conclusions sound. Such use of neuroscience raises questions about moral responsibility and biological determinism—whether our brains cause us to act badly and, if so, whether we can or should be held responsible for these actions.

The popular media have brought these questions to the public's attention. The Public Broadcasting Service television series *Brains on Trial* features actor Alan Alda asking various experts if, and how, neuroscience research might be used in court, and whether scientific advances might challenge traditional notions of moral responsibility.[25] Media articles include discussion about whether neuroscience can prove or disprove that we have free will, and how that might affect our notion of criminal responsibility.[26] A 2014 article discusses how neuroscience evidence increasingly might play a role in decisions about whether one is competent to stand trial, or whether one should be spared a particular punishment because of a brain abnormality.[27] *NBC News* reported on a new technique that uses brain waves to detect whether individuals recognize certain objects, people, or places that they have personally experienced.[28] This test might never make it to U.S. courts, but some law enforcement experts are calling for its increased use as an investigatory tool.[29]

In his charge to the Bioethics Commission, President Obama asked specifically that we address the use of neuroscience in the criminal justice system.[30] In addition, the Bioethics Commission received several public comments from key stakeholders recommending consideration of this topic. For example, the MacArthur Foundation Research Network on Law and Neuroscience sought consideration of the admissibility and weight of neuroscientific evidence in the courtroom, and its application to legal questions about lying and memory, prediction of future criminal behavior, and adolescent conviction and sentencing.[31] The application of neuroscience to the legal system and its potential to alter how we understand free will and blameworthiness are topics of both excitement and concern. The Bioethics Commission includes the topic in this report to examine ethical questions about the use of neuroscience within the legal system and how it can inform our fundamental conception of moral responsibility.

Selected Reports on Neuroscience, Ethics, and Society

The President's Council on Bioethics' 2003 report *Beyond Therapy: Biotechnology and the Pursuit of Happiness* reflected on ethical and societal implications of using emerging biotechnologies for modifying behavior, slowing age-related decline, and altering emotions, among others.

The National Research Council's 2008 report *Emerging Cognitive Neuroscience and Related Technologies* explored neuroscience research areas that might have implications for U.S. national security, including advances in measuring individuals' mental states and intentions, and development of drugs or technologies that can alter human abilities.

The National Research Council's 2009 report *Opportunities in Neuroscience for Future Army Applications* focused on research and initiatives that might improve the cognitive and behavioral performance of soldiers.

The United Kingdom's Royal Society released four *Brain Waves* reports in 2011 and 2012 that reviewed neuroscience and neurotechnology developments and their translation into useful applications, the effect of neuroscience advances on education and lifelong learning, potential military and law enforcement applications arising from neuroscience, and neuroscience's increasing relevance to the law.

The Nuffield Council on Bioethics' 2013 report *Novel Neurotechnologies: Intervening in the Brain* considered the potential benefits and harms of novel neurotechnologies and outlined an ethical framework to guide their future development, regulation, use, and promotion.

Singapore's Bioethics Advisory Committee (BAC) is considering current and developing neuroscience research and its ethical, legal, and societal implications. BAC expects to release its final report and recommendations in the near future.

Sources: The President's Council on Bioethics. (2003, October). *Beyond Therapy: Biotechnology and the Pursuit of Happiness*. Washington, DC: The President's Council on Bioethics. Retrieved February 2, 2015 from https://bioethicsarchive.georgetown.edu/pcbe/reports/beyondtherapy/index.html; Committee on Military and Intelligence Methodology for Emergent Neurophysiological and Cognitive/Neural Research in the Next Two Decades, National Research Council. (2008). *Emerging Cognitive Neuroscience and Related Technologies*. Washington, DC: The National Academies Press. Retrieved February 2, 2015 from http://www.nap.edu/catalog/12177/emerging-cognitive-neuroscience-and-related-technologies; Committee on Opportunities in Neuroscience for Future Army Applications, National Research Council. (2009). *Opportunities in Neuroscience for Future Army Applications*. Washington, DC: The National Academies Press. Retrieved February 2, 2015 from http://www.nap.edu/catalog/12500/opportunities-in-neuroscience-for-future-army-applications; The Royal Society. (2011, January). *Brain Waves Module 1: Neuroscience, Society, and Policy*. London, UK: The Royal Society. Retrieved February 2, 2015 from https://royalsociety.org/policy/projects/brain-waves/society-policy/; The Royal Society. (2011, February). *Brain Waves Module 2: Neuroscience: Implications for Education and Lifelong Learning*. London, UK: The Royal Society. Retrieved February 2, 2015 from https://royalsociety.org/policy/projects/brain-waves/education-lifelong-learning/; The Royal Society. (2012, February). *Brain Waves Module 3: Neuroscience, Conflict, and Security*. London, UK: The Royal Society. Retrieved February 2, 2015 from https://royalsociety.org/ policy/projects/brain-waves/conflict-security/; The Royal Society. (2011, December). *Brain Waves Module 4: Neuroscience and the Law*. London, UK: The Royal Society. Retrieved February 2, 2015 from https://royalsociety.org/policy/projects/brain-waves/responsibility-law/; Nuffield Council on Bioethics. (2013, June). *Novel Neurotechnologies: Intervening in the Brain*. London, UK: Nuffield Council on Bioethics. Retrieved February 2, 2015 from http://nuffieldbioethics.org/project/neurotechnology/; Bioethics Advisory Committee Singapore. (2014). Ethics Guidelines for Human Biomedical Research – Neuroethics [Webpage]. Retrieved February 2, 2015 from http://www.bioethics-singapore.org/index/ activities/current-projects.html.

Background and the Promise of Neuroscience

The human nervous system is an exceptionally intricate structure that regulates every organ and tissue of the body. It enables countless automatic functions that we share with other living creatures, including life-sustaining bodily activities, locomotion, and perception of the environment. In addition, it is the locus of more complex human attributes of language, empathy, judgment, and mood. Philosophers, scientists, and other thoughtful human beings have sought to understand the brain and the basis of consciousness, cognition, and volition for centuries.

The Promise of Neuroscience Research

One of the greatest illustrations of the promise of neuroscience—and a priority for many who study it—is its ability to better the human condition and alleviate suffering. In the original charge to the newly formed National Institutes of Health (NIH) in 1952, Congress defined the purpose of publicly funded biomedical research as follows: "[T]o help provide the practicing physicians of this nation—and of the world—with better means for ameliorating physical suffering and emotional imbalance, for prolonging human life, and for making all the years of that span more useful both to the individual and to society."[32]

NEURONS IN THE BRAIN

Weighing approximately 3 pounds, the human brain comprises 100 billion neurons and over 60,000 miles of connections that facilitate information transmission. A typical neuron links with several thousand others, resulting in 100 trillion synaptic connections, and allowing for more computing power than any machine. Other cells in the brain provide structure and control the environment necessary for optimal brain function, defend against infection, and support tissue remodeling and repair. Electrical signals are passed between neurons through molecules called neurotransmitters. More than 100 neurotransmitters exist, and most drugs that affect brain function work by altering their levels or function.

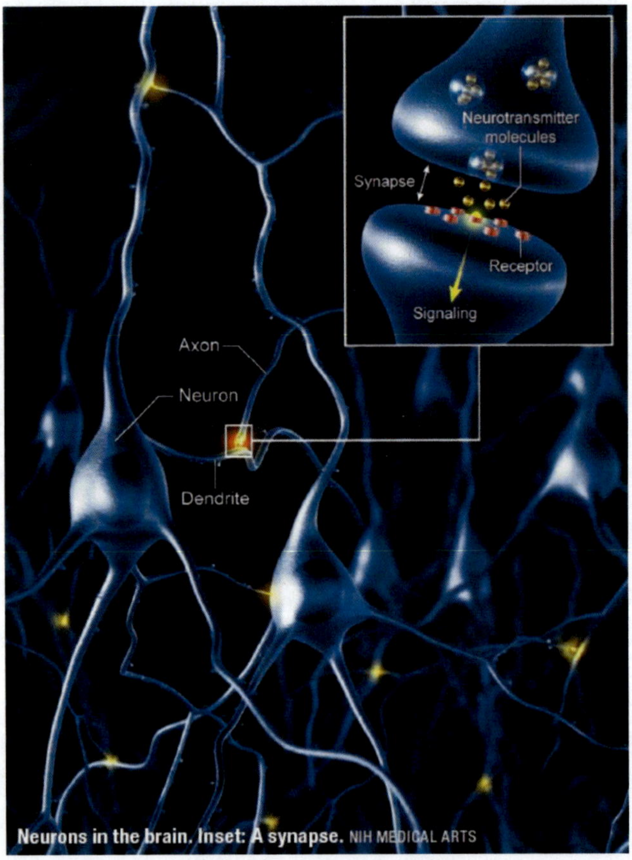

Neurons in the brain. Inset: A synapse. NIH MEDICAL ARTS

Sources: Swaab, D.F. (2010). Developments in Neuroscience. In J.J. Giordano and B. Gordjin (Eds.). *Scientific and Philosophical Perspectives in Neuroethics* (pp. 1-36). New York, NY: Cambridge University Press, p. 3; Verkhratsky, A., and A. Butt. (2013). *Glial Physiology and Pathophysiology*. West Sussex, UK: Blackwell Publishing; Purves, D., et al. (Eds.). (2001). *Neuroscience*, Second Edition. Sunderland, MA: Sinauer Associates.

These founding principles are central to neuroscience. Our nervous system determines our capacity to cope with suffering; adapt to adverse conditions; regulate stress, disappointment, and loss; remember past events; plan for the future; and improve the lives of our loved ones. Neurological disorders also are a substantial source of morbidity and mortality. The impact of neurological disorders and psychiatric conditions can be quantified in disability-adjusted life years (DALYs), which account for years lost to disability and premature death (Table 1). They are common and costly and often have devastating impact. More than one billion individuals globally, and millions in the United States, suffer from neurological disorders, with estimated health system and lost productivity costs of over 760 billion dollars a year in the United States alone.[33] Furthermore, as our population ages, the prevalence of age-related neurologic disorders, including neurodegenerative diseases, such as Alzheimer's disease and Parkinson's disease, will increase.[34]

Historically, clinicians struggled to diagnose and treat neurological disorders. Before the late 1970s, clinical neuroscience focused on relating a patient's symptoms with tissue changes observed at autopsy.[35] As diagnosis of neurological problems improved, their future course could be predicted with varying degrees of accuracy, but in most cases, clinicians were unable to intervene in any meaningful way.[36]

Advances in neuroscience over recent decades led to development of treatments for certain debilitating neurological disorders. For example, multiple sclerosis, the most common cause of nontraumatic neurological impairment among young and working-age adults, was previously untreatable, and the vast majority of affected individuals developed irreversible neurological disability.[37] Over the past two decades, research has yielded more than 10 FDA-approved therapies, dramatically improving the outlook for affected individuals.[38]

Table 1. Global disability-adjusted life years (DALYs) and number of annual deaths for selected neurological disorders and psychiatric conditions in 2010

Disorder	DALYs†	Deaths‡
Low back and neck pain	116,704,000	–
Cerebrovascular diseases (stroke)	102,232,000	5,874,200
Epilepsy	17,429,000	177,600
Multiple sclerosis	1,075,000	18,200
Alzheimer's disease and other dementias	11,349,000	485,700
Parkinson's disease	1,918,000	111,100
Schizophrenia	14,999,000	19,800
Depression	74,264,000	–
Alcohol use disorders	17,644,000	111,100
Drug use disorders	19,994,000	77,600
% of total for 10 selected neurological disorders and psychiatric conditions	**15.2%***	**13.0%**§

Sources: †Murray, C.J.L., et al. (2012). Disability-adjusted life years (DALYs) for 291 diseases and injuries in 21 regions, 1990–2010: A systematic analysis for the Global Burden of Disease Study 2010. *The Lancet*, 380(9859), 2197-2223; ‡Lozano, R., et al. (2012). Global and regional mortality from 235 causes of death for 20 age groups in 1990 and 2010: A systematic analysis for the Global Burden of Disease Study 2010. *The Lancet*, 380(9859), 2095-2128.

*DALYs from 10 selected conditions divided by DALYs from 291 conditions studied; §deaths from 10 selected conditions divided by deaths from 235 causes.

Traumatic brain injury (TBI), a disruption of normal brain function caused by impact from an external force, is the world's leading cause of death and disability among children and young adults.[39] Ongoing clinical trials are testing different strategies to intervene during the acute phase of TBI.[40] Harnessing novel neurotechnologies, including electrical stimulation and neuroimaging techniques, might provide new or improved ways to diagnose and treat TBI.[41] Additionally, advances in neuroimaging might aid in assessing awareness among individuals in a minimally conscious state caused by brain injury.[42]

An estimated 50 million individuals worldwide have epilepsy, a neurological disorder that is a major cause of morbidity and mortality and can lead to stigmatization.[43] Contemporary neuroscience led to development of several dozen effective drugs to treat epilepsy. In certain cases, neuroimaging that measures electrical activity of specific brain regions enables clinicians to identify culprit areas of the brain, potentially leading to new treatment options.[44]

Deep Brain Stimulation

Deep brain stimulation (DBS) is a surgical intervention that delivers rapidly fluctuating electric current to deep brain structures. The exact mechanism of DBS action is unknown, but it remains a viable treatment option for certain disorders.

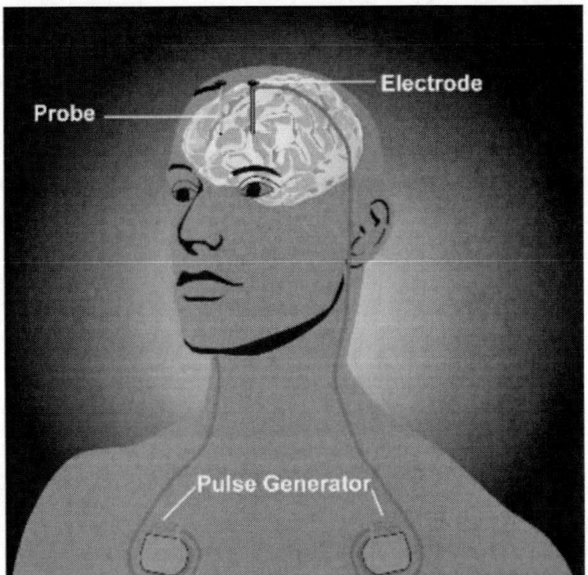

Diagram of Implanted DBS Apparatus from: National Institute of Mental Health, National Institutes of Health, Department of Health and Human Services.

DBS is approved by the U.S. Food and Drug Administration for treating Parkinson's disease, other movement disorders, and treatment-resistant obsessive-compulsive disorder. Research indicates that DBS might be effective in treating neurological disorders and psychiatric conditions, such as depression, epilepsy, and chronic pain.

DBS also has potential to alter individuals' behavior. Because of the invasive nature of the procedure, clinicians and patients consider DBS only after failure of nonsurgical interventions.

Sources: Perlmutter, J.S., and J.W. Mink. (2006). Deep brain stimulation. *Annual Review of Neuroscience*, 29, 229-257; Kern, D.S., and R. Kumar. (2007). Deep brain stimulation. *The Neurologist*, 13(5), 237-252; Shah, R.S., et al. (2010). Deep brain stimulation: Technology at the cutting edge. *Journal of Clinical Neurology*, 6(4), 167-182; Chen, X.L., et al. (2013). Deep brain stimulation. *Interventional Neurology*, 1(3-4), 200-212; Murrow, R.W. (2014). Penfield's prediction: A mechanism for deep brain stimulation. *Frontiers in Neurology*, 5(213), 1-8.

Many neurodegenerative diseases—including Alzheimer's disease, Parkinson's disease, amyotrophic lateral sclerosis (Lou Gehrig's disease), chronic traumatic encephalopathy (resulting from repeated brain trauma), and Huntington's disease, among others—are associated with accumulation of clumped proteins in degenerating nerve cells.[45] Evidence indicates that misfolding of culprit proteins into incorrect shapes might be responsible for the degenerative process in each disease.[46] This misfolding might be genetic or triggered by injury, such as by trauma or infection. Measuring and imaging these abnormal proteins directly and noninvasively in patients is now becoming possible. These advances create a remarkable opportunity to develop effective therapies for neurodegenerative diseases and to monitor their effectiveness in the living brain.

Technological developments in neuroscience offer new hope for individuals with debilitating neurological disorders. For example, deep brain stimulation (DBS) can enable motor function by eliminating disabling tremors, improving the lives of many thousands of individuals with Parkinson's disease.[47] In addition, DBS offers promise for other disorders, including psychiatric conditions.[48]

In recent years the practice of neuroscience has improved and changed dramatically, fueled by a proliferation of enabling technologies that are transforming our understanding of brain function in health and disease. These technologies include neuroimaging, molecular and cellular neuroscience, medicinal chemistry, and bioinformatics, among others. Sophisticated techniques and technologies have improved prospects in diagnosis, prevention, and treatment of neurological disorders and psychiatric conditions. They have also been central to contributing to a deeper understanding of the human brain, consciousness, behaviors, motivations, intentions, and our sense of self.

Recent Advances in Neuroscience

In 2013, scientists developed a method to preserve the integrity of biological tissue while improving its visibility under the microscope. The method replaces dense, irregularly arranged lipid molecules present in normal biological tissue with a hydrogel that allows light and other molecules to pass through easily. The resulting tissue-hydrogel hybrid is transparent, facilitating visualization of large networks of neurons and their connections in the whole, intact brain.

In 2012 and 2013, two teams of scientists demonstrated methods to create a false memory association in mice. Using different techniques and tools, both teams of researchers activated neurons that corresponded to the memory of a foot shock. The scientists were able to reinstate the fear of a foot shock even if the mouse was not placed in the original context in which the foot shock occurred.

In 2011, scientists developed a decoder able to identify and reconstruct moving images viewed by a human participant. Researchers first presented participants with a movie clip and simultaneously measured their brain activity through functional magnetic resonance imaging (fMRI). The decoder, a model of the brain's visual system, used fMRI data from each individual to produce a primitive reconstruction of the movies seen by the participant.

Sources: Chung, K., et al. (2013). Structural and molecular interrogation of intact biological systems. *Nature*, 497(7449), 332-337; Garner, A.R., et al. (2012). Generation of a synthetic memory trace. *Science*, 335(6075), 1513-1516; Liu, X., et al. (2012). Optogenetic stimulation of a hippocampal engram activates fear memory recall. *Nature*, 484(7394), 381-385; Ramirez, S., et al. (2013). Creating a false memory in the hippocampus. *Science*, 341(6144), 387-391; Nishimoto, S., et al. (2011). Reconstructing visual experiences from brain activity evoked by natural movies. *Current Biology*, 21(19), 1641-1646.

The BRAIN Initiative

The BRAIN Initiative, a White House Grand Challenge, represents substantial federal investment in neuroscience research and a unique opportunity to advance the fruits of neuroscience and its potential to improve human health.

The BRAIN Initiative supports public and private research to develop tools and technologies to understand the human brain. Advances in the ability to map the anatomy and activity of neurons and their connections in different regions in the brain create an opportunity to decipher how groups of neurons and their connections function normally and in disease states. These networks are critical to understanding neurodegenerative disorders. Additional potential exists to understand psychiatric conditions, including depression, schizophrenia, anxiety, post-traumatic stress disorder (PTSD), and addiction, among others. Understanding networks of neurons to decipher the signature causes of these disorders can help scientists develop more effective diagnostic tools, treatments, prevention, and cures.

Contemporary neuroscience is a rapidly growing, multidisciplinary field, and neuroscientists come from varied academic backgrounds and subfields. Even in such a vast and varied field, we can agree on the value of both the intermediate goal of advancing human knowledge through neuroscience research and the ultimate goal of alleviating human suffering and ameliorating neurological disorders that burden individuals. The BRAIN Initiative, pursued with acumen, persistence, and adequate funding for multidisciplinary teams of researchers, has the potential to help us realize the promise of neuroscience. It supports the development of innovative technologies to improve our understanding of brain function with the ultimate goal to better characterize and treat neurological disorders, including Alzheimer's and Parkinson's diseases, depression, PTSD, and TBI.[49] Although more than innovative technology alone is needed to realize the ultimate goal of the BRAIN Initiative, knowledge gained through the use of these new technologies could be key to many diagnostic tools, treatments, preventions, and cures for these and other debilitating neurological disorders.

About this Report

In July 2013, President Obama charged the Bioethics Commission to consider the ethics of neuroscience research and application of neuroscience research findings as part of the BRAIN Initiative.[50] *Gray Matters*, Vol. 1, released in May 2014, describes many options to achieve integration of ethics with neuroscience research, arguing that a single approach is neither sufficient nor appropriate for all contexts. Integrating ethics and neuroscience is complex: it requires understanding and addressing the unique needs and contexts of individual researchers and institutions. Whatever approach or set of approaches are chosen to integrate neuroscience and ethics, funding and resources are required. For example, scholars have engaged in extensive discussion about one historical model for integration, the NIH's and U.S. Department of Energy's Human Genome Project Ethical, Legal, and Social Implications Research Program (HGP ELSI Program). They generally agree that the funding provision for the HGP ELSI Program, which accompanied funds for scientific research, was a positive aspect of the program. The Bioethics Commission agreed and recommended in *Gray Matters*, Vol. 1 that funding be allocated for ethics integration in neuroscience.[51]

This report considers in greater detail some of the ethical and societal implications of neuroscience research and its varied applications—implications that integrated ethics and research systems are well-positioned to address. Through deliberation at seven public meetings since receiving the President's charge, in addition to two relevant meetings undertaken as part of earlier work, and hearing from over 60 experts, the Bioethics Commission identified three controversial topics that exemplify some of the primary ethical considerations at the intersection of neuroscience and ethics. In this report, the Bioethics Commission explores these topics: cognitive enhancement and beyond, capacity and the consent process, and neuroscience and the legal system. The following sections examine these topics, clarify the current state of the scientific landscape, identify common ground among differing viewpoints, offer an in-depth analysis of ethical considerations, and make concrete recommendations for relevant stakeholders.

SECTION 2. COGNITIVE ENHANCEMENT AND BEYOND

Scientific investigations of the brain and nervous system point the way to new techniques and technologies to modify human neural functioning. Pharmaceuticals, brain stimulation devices, and brain training tools are just some of the neural modification modalities that are in use, under study, or anticipated. Scholars debate the meaning of the term "cognitive enhancement." Generally, cognitive enhancement describes measures for expanding or augmenting the human capacity to think, feel, react, and remember, potentially "beyond the species-typical level or statistically-normal range of functioning."[52] Some parties in the debate welcome novel cognitive enhancements as means to human betterment.[53] Others reject the use of biomedical innovations intended to push cognition beyond typical human functioning as threats to moral agency and dignity.[54] What might happen, scholars ask, to traditional understandings of free will, moral responsibility, and virtue if science makes significant advances in the ability to technologically control the mind? In this report, the Bioethics Commission seeks to engage the public in discussion that centers on this important cognitive enhancement debate and moves beyond it to assess a wider array of interventions, technologies, behaviors, and environmental conditions that can affect the functioning of the human brain and nervous system. We use the term "neural modifiers" to refer to this wider array of mechanisms of brain and nervous system change.

Our brains and nervous systems constantly change in response to ordinary daily environmental stimuli, including education, meditation, physical activity, sleep, and diet, among others. Neuroscience helps us better understand the mechanisms of neural change and design novel interventions to alter our brains and nervous systems for a variety of purposes deemed beneficial. Bold new life-changing treatments that improve cognitive, nervous system, and motor function are in use already and more are on the horizon. Researchers seek new approaches to (1) *maintain or improve* neural health and cognitive function within typical or statistically normal ranges; (2) *treat* disease, deficiency, injury, impairment, or disorder (referred to as "neurological disorders") to achieve or restore typical or statistically normal functioning; and (3) *expand or augment* function above typical or statistically normal ranges. In delineating these neural modification objectives, the Bioethics Commission is mindful that they are not always sharply distinguishable. Drawing clear lines between *maintaining or improving* function within normal ranges on one hand and *expanding or augmenting* on the other, or between *treating* as therapy on one hand and *expanding or augmenting* as enhancement on the other, can be difficult in both theory and practice. Moreover, it bears special emphasis that what is "normal" forms a bell-shaped curve that encompasses a range of individual differences, and ethical analysis should not uncritically embrace conventional understandings of normal as ideals of healthy bodies and minds.

Our personal, professional, and societal intentions—whether to maintain or improve, treat, or expand and augment the brain and nervous system—are subject to moral evaluation, as are our ultimate actions and omissions. Altering the brain and nervous system is not inherently ethical or unethical. We can imagine that some neural modifications would violate shared values, whereas others would not. For example, individuals report feeling comfortable with the use of neural modifiers for therapeutic purposes, but are less comfortable about healthy individuals using such products.[55] Additionally, although one might view use of prescription medication as medically necessary to treat post-traumatic stress disorder (PTSD), one also might view the overuse of some of the same drugs, and the resulting memory alteration, as controversial threats to personal identity, among other concerns.[56]

Ethical assessment of neural modification requires consideration of who is choosing the modifier, what is being chosen, what its purposes are, who stands to benefit, and who might be harmed. For example, deep brain stimulation (DBS) is currently used to treat motor disorders, such as Parkinson's disease, and is under investigation for certain psychiatric conditions. If demonstrated to be effective for problems such as depression, addiction, or overeating associated with obesity, DBS could be ethically acceptable if freely chosen by a fully informed adult. In contrast, the intervention would be ethically problematic if a person is coerced into undergoing DBS by a clinician, researcher, or government authority.[57] A pharmacological intervention, such as modafinil, might be used ethically to treat a disorder of sleep and behavior, such as narcolepsy, but might be considered ethically problematic when used recreationally or to further improve a healthy child's performance in school.[58] Interventions that are available to improve the health of everyone might pass muster, yet those reserved for economically privileged groups would affront fairness and equity.

In this time of rapidly advancing neuroscience, members of the public must be well-informed, both to make educated practical decisions about personal health and wellbeing, and to participate in collective deliberation and decision making about societal applications of novel techniques and technologies relating to the brain and nervous system. The public should be encouraged and empowered to be ethically thoughtful about where responsibility rests for behaviors and conditions that adversely affect neural health, wellbeing, and the capacity for a good life. Society must be prepared to assess novel neural modifiers as they are developed, to ensure that benefits justify risks, and to assure that relevant ethical concerns are considered, including freedom from unwarranted coercion, fair access, and the preservation of moral capacities that define us as human. The Bioethics Commission seeks in this report to describe the landscape of neural modification—including but not limited to cognitive enhancement—dispel common misconceptions, help educate the broader public, and guide an ethical debate that should be based on reliable empirical and scientific evidence.

Goals and Purposes of Neural Modification

As scientific knowledge of the human brain and nervous system expands, so too will our ability to intentionally modify our neural functioning. Interventions that change neural function exist—and more are anticipated—enabled by developments in neuroscience related to memory, learning, intelligence, cognition, and motor and sensory function. As outlined above, goals and purposes of neural modification include: (1) *maintaining or improving* neural health and cognitive function within the range of typical or statistically normal human functioning, (2) *treating* neurological disorders, and (3) *expanding or augmenting* neural function. Against a background of neuroscientific discovery, neural modification with these objectives has emerged as a principal topic for bioethics to consider.[59] Neural modification approaches that go beyond ordinary, low-technology activities into the realm of pharmaceuticals and invasive surgery are of greatest ethical concern due to their heightened potential risk, and possibility for more drastic or irreversible change. We are less concerned about caffeinated coffee, healthy sleep habits, and good nutrition, for example, than we are about use of stimulant drugs and DBS implantation, especially where the intervention serves purposes other than maintaining or improving health or therapeutic treatment of neurological disorders.

Maintaining or Improving Neural Health and Cognitive Function

Maintaining or improving neural health and cognitive function within typical or statistically normal ranges is often a primary goal of modifying the brain and nervous system (Figure 1). Interventions and methods that improve individual and public health and social and environmental conditions can promote and maintain brain and nervous system function.[60] Many ordinary daily activities can qualify as neural modifiers, and many are presumed ethical because they seem natural and safe, and contribute to a good life in the modern world. For example, good nutrition is essential for optimal cognitive function, and many commentators focus on the role of specific nutrients, such as omega-3 fatty acids, in developing cognitive function.[61] Although the effect of omega-3 fatty acids on cognitive development is debated, some evidence indicates that supplementing this nutrient among selected populations, such as infants, might be beneficial for several measures of cognitive development.[62]

Modification of other health behaviors, such as sleep and exercise, also can contribute to maintaining cognitive function. Public health interventions, such as lead abatement programs, can remove impediments to optimal cognitive functioning on the population level as well.[63]

At the individual level, activities such as getting a full night's sleep, exercising, and maintaining a nutritious diet contribute to brain health and function. Core literacy and numeracy skills and devices, such as computers, calculators, and smartphones, increase our ability to access and process information, concentrate on or attend to information, and communicate with others. Even the caffeine in morning coffee and afternoon tea might qualify as neural modification for the boost to alertness they provide. A cleaner and more pollution- and toxin-free natural environment, along with healthy residential living and working conditions, can contribute to sound neural function. Achieving some of these improved conditions for all comes with a considerable economic price. Yet, as the Bioethics Commission has previously reported, deliberate neglect of the societal background conditions needed for public health is costly, not just in economic terms; neglect of these well-understood improvements in favor of focusing exclusively on new and emerging intervention technologies runs afoul of principles of public beneficence and justice.[64] With regard to neural health, the relevant background conditions are numerous and therefore provide a range of options worth considering alongside and in addition to investments in neuroscience; these include investments in basic literacy, numeracy, science education, and ethical literacy, as well as environmental protection, and residential and occupational safety.

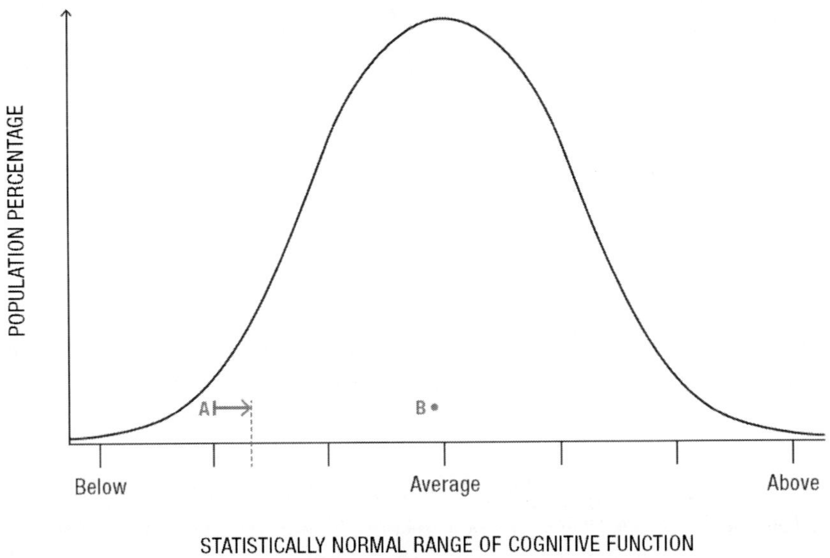

A) An individual with low (but statistically normal) cognitive function uses a neural modifier to improve cognition

B) An individual takes a dietary supplement to maintain healthy cognitive function.

Figure 1. Maintaining or Improving Neural Health and Cognitive Function within Typical or Statistically Normal Ranges.

> "[C]onsumers may not take actions that we know are useful...following a healthy diet, exercise, learning new things, all of which...people might be less likely to do if they think there's a pill out there that solves the problem."
>
> Viswanathan, S., Assistant Director, Division of Advertising Practices, Bureau of Consumer Protection, U.S. Federal Trade Commission, (2014). Ethical Responsibilities of Direct-to-Consumer Neuroscience Companies. Presentation to the Bioethics Commission, November 5. Retrieved February 3, 2015, from http://bioethics.gov/node/4320.

More controversially, some scholars have suggested that drugs or devices might be developed that can improve cognitive function within statistically normal ranges; however, this is a matter of considerable debate. If and when they become available, these new higher-technology strategies to improve neural function within the normal range might be more controversial than existing methods. To draw an analogy, recombinant human growth hormone is a safe and readily available drug, approved by the U.S. Food and Drug Administration (FDA) to treat growth hormone deficiency and other growth disorders.[65] Growth hormone also has been used to increase the height of children substantially shorter than the mean for their age and sex, but who are nonetheless "normal" and do not have a diagnosable disorder.[66] The use of this drug has raised concerns about whether it is morally acceptable to use drugs to alter the natural, non-diseased state. Other debates around this issue include whether use of human growth hormone for this purpose should be funded, and by whom.[67]

A similar debate exists in the cognitive enhancement literature. On the one hand, some scholars argue that, if drugs and devices become available that can improve cognitive or other neural function within statistically normal ranges, there is nothing *per se* unethical about their use. These scholars suggest that novel, high-technology strategies to improve cognitive performance are not inherently different from older, more familiar methods, such as education and calculators. For these scholars, novel interventions do not raise new normative questions although they might raise concerns about safety, distributive justice, or unintended consequences.[68] On the other hand, other scholars contend that drugs and devices used for this purpose reflect a desire to control or master human nature. This desire for control might erode our appreciation for natural human powers and achievements.[69] Neural modification that improves functions relative to a standard distribution is controversial. Some scholars assert that if access to neural modification is universal, gradually increasing average capabilities might fail to improve the relative wellbeing of the least well off.[70] Other scholars view gradual improvement in average neural functioning as a social good worth pursuing.[71] Additional ethical concerns about gradually increasing mean levels of neural functioning include the reduction of diversity (e.g., if neural modification is targeted iteratively to those on the lower end of the normal distribution), and the specter of unintended consequences.

As we reflect on the potential of the BRAIN Initiative and other neuroscience advances to alleviate human suffering, it is worth emphasizing that, if we as a society and individuals have the will, we already have some practical and ethical ways to make substantial improvements in human neural functioning, even before the neuroscientific advances enable us to do more. Correspondingly, the added knowledge gained by such endeavors as the

BRAIN Initiative will need to be supplemented by the social will to invest the resources that are needed to put that knowledge to use in bettering the human condition.

Treating Neurological Disorders

A major goal of modifying the brain or nervous system is to treat neurological disorders. Illness, mishap, and aging can result in considerable impairment. Some neural modification is aimed at enabling persons to think, feel, or remember substantially better than they otherwise could.

In the United States and worldwide, the burden of neurological disorders is high and projected to increase considerably in future years. Neurological disorders and their sequelae are estimated to affect as many as a billion individuals globally, including millions in the United States.[72] Furthermore, these disorders occur among all age groups and geographical regions.[73]

For example, Alzheimer's disease can cause severe memory loss and personality changes. It is the most common neurodegenerative disease in the United States and affects an estimated 5.2 million individuals.[74] Parkinson's disease impairs motor function and in certain cases can cause dementia. It affects an estimated one million Americans, even striking those who are young.[75] Congenital epilepsies and cerebral palsy can be so severe as to interfere with childhood learning and adult employment. Epilepsy, the fourth most common neurological disorder, affects an estimated 2.2 million U.S. residents.[76] Cerebral palsy, the most common childhood motor disability, affects an estimated one in 323 U.S. children.[77] Autism spectrum diagnoses have increased dramatically during the past several decades, with an estimated one in 68 children being affected.[78] Multiple sclerosis, an autoimmune disease of the nervous system and a common cause of neurologic disability among young adults, also has increased in frequency, especially among women.[79] Automobile crash-related injuries, workplace injuries, combat injuries, and sports injuries leave many individuals without normal brain function, limbs, sight, or hearing.[80] Traumatic brain injuries (TBI), most often caused by car crashes, sports injuries, injuries from military combat, and falls, are the leading cause of death and disability among persons aged 1 to 44 years.[81] Approximately 5.3 million suffer from long-term disability as a result of TBI.[82]

In addition, millions of persons living in the United States are affected by psychiatric conditions, including schizophrenia, bipolar disorder, personality disorder, and major depression. Approximately one in four adults (61.5 million) experiences mental illness in any given year.[83] Approximately 13.6 million individuals live with an ongoing serious mental illness.[84] These disorders can involve disorganized thinking, emotional pain and suffering, extreme emotional dysregulation, and limited self-control, all of which can affect pediatric populations as well as adults. Many serious addictions also are understood today as treatable mental and behavioral health conditions.

The financial cost to society of treating mental and behavioral health conditions is high. For example, the United States spends 113 billion dollars annually on mental health treatment, and serious mental illness is estimated to cost 193 billion dollars in lost earnings per year.[85] But even greater, although in some respects incalculable, are the personal and social costs of caring for those affected by these conditions.

Helping affected individuals to be well again or for the first time is vastly appealing as a public health priority and an ethical imperative of beneficence and justice. As discussed above, one of the primary goals of neuroscience is to relieve human suffering and better the human condition.

Expanding or Augmenting Neural Function

A third major goal of changing the brain and nervous system is to expand or augment cognitive and motor function. The literature refers to many purposes of neural modification to expand or augment function, including cosmetic enhancement, enhancement for the sake of competitive advantage, moral enhancement, and transhumanism (a movement seeking to use technology to radically enhance the human condition). What counts as expanding or augmenting a human trait is contested and subject to debate. Some groups are excited about the possibility that society could be improved by raising the baseline of what is considered within normal human limits. At its most fanciful, this goal is associated with the idea of a future in which human beings will be equipped with functioning more advanced than the typical or statistical norm, such as unyielding stamina or perfect recall by way of bionic men and women, cybernetic organisms, and transhumanism.[86] In a more realistic form, the goal of extending or augmenting neural function is associated with modest cognitive enhancement.

The term "cognitive enhancement" is defined in different ways by the scholars who discuss it, but most uses refer to interventions such as pharmaceuticals, technological devices, and surgeries that improve abilities to think, feel, and remember.[87] Cognitive enhancement is controversial as applied to interventions made available to healthy individuals functioning within normal limits. For example, using psychedelic drugs, including psilocybin and lysergic acid diethylamide (LSD) to achieve what is believed to be spiritual transcendence has been controversial, and criminal law prohibits their use.[88] Ethical concerns about justice and fairness, among others, are raised by prescribing stimulant medications like methylphenidate (Ritalin®) to enable a healthy child without learning disabilities to achieve a higher score on a standardized test.[89] But concerns also are raised by using propranolol to relieve stage fright among performers and to dampen the traumatic memories of survivors of war or sexual assault. These concerns include the ethics of performing better through chemistry, the potential for over-medicalization, and "subsequent exploitation by the pharmaceutical industry."[90] Further, when we consider altering our memories, we trigger concerns at the core of defining one's self. Individuals' memories guide their narrative identities, and modifying them might alter what individuals believe to be true about themselves.[91] Enhancements praised by some observers as improvements are condemned by others as assaults on human dignity, autonomy, and moral virtue.[92]

Intentional modifications of the function of the brain and nervous system, including those debated under the rubric of "cognitive enhancement" raise important ethical questions with implications for clinical care, biomedical research, public policy, and the law. Pharmaceuticals are among the most discussed neural enhancers, but reliable data on the actual prevalence of use of pharmaceutical interventions for cognitive expansion or augmentation are elusive. Few data exist on the prevalence of use of novel pharmaceuticals to boost cognition. The most recent National Survey on Drug Use and Health from the Substance Abuse and Mental Health Services Administration reported that 6.4 percent of college students aged 18–22 reported using combined levo- and dextroamphetamines (Adderall®) non-medically.[93] Survey participants reported using these drugs to improve concentration, focus studying, and increase alertness.[94] An online poll of journal readers conducted by *Nature* found that one in five of 1,400 respondents had used Ritalin®, Provigil®, or beta-blockers for non-medical purposes (stimulating focus, concentration, or memory).[95] These results must be interpreted with caution because they were drawn from a small, nonrepresentative sample and are limited by selection bias; however the data demonstrate attempts by healthy individuals to engage in pharmaceutical forms of cognitive enhancement.[96]

Scholars have raised concerns about the available data on the prevalence of cognitive enhancement.[97] Individuals generally are reluctant to admit using prescription drugs for non-medical reasons, making it difficult to estimate prevalence.[98] Published surveys focus on specific populations (e.g., college students) rather than the general population.[99] This limited focus can undermine claims about the widespread use of stimulants for cognitive enhancement.[100] Surveys often do not capture accurately whether the drugs are being used for cognitive enhancement or other purposes. For example, many surveys ask respondents about non-medical use, but do not distinguish between non-medical use for enhancement and recreational purposes.[101] Inadequate data on the prevalence of use of novel forms of cognitive enhancement can lead to under- or overestimates. Generalizing findings from selected groups more likely to use cognitive enhancing drugs can lead to overestimation of prevalance and a misleading sense of urgency regarding the need for regulatory frameworks and policies.[102] Better evidence of prevalence is essential for a careful and considered assessment of the ethical challenges raised by novel forms of cognitive enhancement. Clearly, as neuroscience advances, individuals will continue to experiment with new ways to enhance their cognition, and we will learn more about the benefits and risks of different modalities of enhancement.

Neuroscience research demonstrates that some drugs and brain stimulation devices can have modest enhancing effects on some cognitive abilities in healthy individuals under certain conditions.[103] However, the size of these effects and their generalizability to real-world (rather than laboratory) settings remain uncertain.[104] These include the stimulant drugs Ritalin® and Adderall®, used in the treatment of attention deficit hyperactivity disorder (ADHD); Provigil®, prescribed for disorders of excessive sleep; and such methods as transcranial magnetic stimulation and transcranial direct current stimulation (tDCS). For example, one review of Provigil® and Ritalin® use for cognitive enhancement states that "expectations regarding the effectiveness of these drugs exceed their actual effects."[105]

This observation illustrates one problematic consequence of hype: Although the actual enhancing effects of these interventions might be limited and can vary among individuals, widespread discussion and debate around them can contribute to unfounded and inflated expectations. In addition, a lack of discussion about unrealistic expectations of benefits and actual risks of side effects can result in more harm to individuals. Novel and still largely experimental modalities of cognitive enhancement raise ethical considerations not relevant when considering more conventional modes because we know and understand less about their potential benefits, if any; their potential risks, including long-term effects; and the potential societal impact of their widespread use likewise are underexplored.

Individuals use many of these interventions therapeutically, and they have well-documented safety profiles in the medical context. However, in the context of healthy individuals, a lack of reliable long-term data on the use of pharmaceuticals for nontherapeutic purposes means that the potential risks are unknown. Some potential risks, such as the risk of dependence, are of particular concern.[106] The risk-to-benefit profile for healthy individuals using pharmaceutical cognitive enhancement is very different than it is for individuals using these interventions for therapeutic purposes, especially given the lack of evidence on the cognitive enhancing effects.

"[W]e've come a long way in understanding the problems...with Ritalin and amphetamine use for people with specific disorders. But as with every drug there is the potential for abuse, and there's the potential for diversion."

Delany, P.J., Director, Center for Behavioral Health Statistics and Quality Substance Abuse and Mental Health Services Administration. (2014). Ethical Considerations in Cognitive Enhancement. Presentation to the Bioethics Commission, August 20. Retrieved February 4, 2015 from http:// bioethics.gov/node/4001.

Even less is known about the risks and long-term effects of novel cognitive enhancers among children and adolescents. Unknown long-term risk is of concern, especially in the context of children's developing brains.[107] Both the use of prescription drugs to treat ADHD among children and the use of prescription drugs for cognitive enhancement among children are increasing.[108] Four percent of 12th graders in the United States reported use of Adderall® and 2 percent use of Ritalin® when asked what amphetamines they have taken during the previous year without a doctor's orders.[109] Pediatric use without a clinician's supervision is concerning because ADHD medications can cause rare but serious cardiovascular side effects, especially in children with a family history or known risk factors.[110] Potential also exists for abuse or dependency.[111] Research with animals demonstrates that Ritalin® can induce changes in neuronal morphology, and can cause behavioral changes in adolescent mice and rats, that persist through adulthood.[112] A position paper endorsed by the American Academy of Neurology, Child Neurology Society, and American Neurological Association recommends that physicians refrain from prescribing drugs for nontherapeutic purposes among children.[113] More research is needed about the neurodevelopmental and behavioral implications of stimulant drug use in children and adolescents.[114]

"[P]hysicians report getting requests for enhancement each week.... But the only policy guidance that they have been given is that they have neither a moral nor legal obligation to prescribe the enhancements. Nor a moral or legal prohibition against prescribing enhancements. So what that does is it leaves the physicians to decide for themselves. But physicians don't have the kind of training that they need to make those decisions. And they've already expressed their ambivalence about being the gatekeepers for this issue."

Reiner, P., Professor, National Core for Neuroethics, Kinsmen Laboratory of Neurological Research and Brain Research Centre, Department of Psychiatry, University of British Columbia. (2014). Cognitive Enhancement. Presentation to the Bioethics Commission, August 20. Retrieved February 4, 2015 from http://bioethics.gov/ node/4001.

Individuals also can purchase some neural modifiers directly from companies without a medical intermediary, for example, over-the-counter dietary supplements or tDCS devices. However, other modifiers (e.g., prescription medications) require a treating clinician. Although consumers generally are responsible for learning about and taking responsibility for any risks involved in using an intervention purchased without a health care professional's oversight, when clinicians are involved, they have a responsibility to consider patients' best interests and warn them about potential risks of taking prescription medication—especially for non-medical use.

No comprehensive or agreed-upon guidance or best-practices document is available to advise stakeholders (including clinicians, employers, parents, educators, and patients, among others) about potential benefits and risks inherent in using a neural modification intervention. Educators, parents, clinicians, and employers in highly competitive fields would benefit from having expert, evidence-based advice when faced with student or employee use of neural modification interventions.

> "[T]he educational component [of understanding the effects of cognitive enhancement] is huge. In high school, I think health classes should include education to the potential users or the current users about the potential detriment and the risks that they are engaging in."
>
> Eaton, M., Former Senior Research Scholar, Stanford University Center for Biomedical Ethics, Former Lecturer, Stanford University Graduate School of Business and School of Medicine. (2014). Ethical Responsibilities of DTC Neuroscience Companies. Presentation to the Bioethics Commission, November 5. Retrieved February 4, 2015 from http://bioethics.gov/node/4320.

Novel means of enhancing cognition that rely on advances in neuroscience, although still in their infancy, are likely to become more effective in the future. Neuroscience research has much to contribute to our understanding of whether these interventions are effective at enhancing cognition and the level of risk they impose for healthy individuals, including children.[115] Thus, now is an opportune time to engage the public and all relevant stakeholders in a discussion about the ethics of novel neural modifiers.

Ethical Analysis

Modifying the brain and nervous system is not inherently ethically problematic. Individuals use a wide range of substances, processes, and interventions to modify the brain and nervous system, including high-quality nutrition, meditation, education, drugs, and devices. Society must evaluate the ethical concerns of specific means of neural modification individually, including those labeled cognitive enhancements, to determine whether and why they are potentially problematic. Scholars characterize ethical issues raised by cognitive enhancement into multiple clusters, including "freedom [and] autonomy, health [and] safety, fairness [and] equity, societal disruption, and human dignity."[116] The debates about cognitive enhancement include many of the ethical concerns raised by neural modification more generally, including the importance of facilitating healthy development and wellbeing; respecting moral agency; informed consent to medical procedures and research; minimization of risk; public education and deliberation; equity and access across all demographic groups; and the reduction of disadvantage, suffering, and stigma associated with neurological disorders.

Benefiting Individuals and Society

Principles of beneficence and non-maleficence require taking steps to promote the health and wellbeing of oneself and others and avoiding harm. With these ideals in mind, society has a responsibility to consider the potential benefits and risks of neural modifiers. Safe and effective neural modification that can alleviate suffering and the felt burdens of human impairment are generally considered morally acceptable.[117] In fact, development and use of neural modifiers to maintain or improve health or treat disease represents one of the primary goals of neuroscience research and advances both individual and public beneficence.

Neural modifiers that maintain or improve health, or treat neurological disorders might not be universally applicable to all members of society, but can be beneficial to individuals who need an intervention. For example, correcting a vitamin B12 deficiency in an individual might be effective in maintaining healthy cognitive function and staving off cognitive decline.[118] Or, in a clinical context, a psychiatrist might prescribe the drug clonazepam (Klonopin®) for a patient suddenly experiencing severe panic attacks after a home invasion.[119] If the drug reduces the frequency and severity of panic symptoms, the patient might return to the quality of work and home life enjoyed before symptom onset.[120] Nothing is inherently unethical about voluntarily taking a dietary supplement known to be safe or a prescription medication shown to be safe and effective by adequate scientific research to maintain or improve health or treat a disease or disorder.[121]

Neural modifiers that improve cognitive ability also offer considerable instrumental benefits. Cognitive abilities can influence important outcomes for individual lives, including success at work, earning potential, likelihood of experiencing social and economic difficulties, and overall health.[122] On a societal level, widespread improvements in cognitive function might produce collective benefits, such as economic gains or improved safety from error reductions in high-risk professions and the military.[123] Certain neural modifiers, such as education and those that treat or prevent disease, could be morally required in societies capable of delivering them because of their potential to advance both individual and public beneficence. For example, in the United States, where access to elementary and secondary education is a legal mandate, enhancing cognitive capacities through schooling is a priority.

Neural modifiers can thus provide substantial individual and societal benefits through interventions that nurture healthy brain development, counteract the effects of impediments to development, and promote optimal functioning.

Justice and Fairness

Concerns about justice and fairness related to neural modifiers that enhance cognition and other functions of the brain and nervous system arise in two distinct ways. First, an individual with more—or enhanced—cognitive abilities, for example, might have an advantage relative to others; in this sense, cognitive ability is a positional good, in that it confers an advantage on some individuals only if others do not have the same good. Cognitive enhancement raises the concern that those who have access will gain an unfair competitive advantage over those who do not.[124] If safe and effective novel forms of neural modification are available only to those who are already advantaged (by wealth or social capital), limited availability might exacerbate existing inequalities.

Justice and fairness requires not only equitable distribution of the benefits of neural modifiers, but it also requires attention to the distribution of their burdens and risks. For example, early study and use of neural modifiers might find that they are effective in the short term, but cause negative consequences in the long term. The burdens and risks of understudied neural modifiers must not fall unfairly on certain groups or individuals.

Second, neural modifiers thought to alter cognitive and other neural functions can offer nonpositional benefits (i.e., benefits that are inherently valuable, not because they provide a competitive advantage over others). For example, having a higher earning capacity is beneficial to individuals not (or not only) because it gives them an advantage over others, but because it provides them with means to secure better living conditions and a greater range of opportunities. Similarly, knowledge is considered a good in itself, and access to more knowledge can be inherently valuable. Here, the concern about justice is not whether access to the means for elevated brain and nervous system function confers unfair advantages, but rather whether the distribution of safe and effective neural modifiers can promote justice by providing individuals with a greater range of opportunities and enabling them to participate more fully in society.[125]

The nonpositional individual and societal benefits of neural modification support pursuing modifications collectively, rather than limiting access to a privileged few. Neuroscience research on the effects of novel neural modifiers can contribute to our understanding of how these interventions can be distributed justly. For example, evidence demonstrates that the effects of certain pharmaceutical cognitive enhancements depend on baseline cognitive functioning. Individuals with lower levels of baseline functioning appear to experience a greater improvement than those at a higher baseline level.[126] If these results are borne out by further research, cognitive enhancement interventions could be used to reduce inequities between the cognitively advantaged and disadvantaged, for example, by reducing gaps in educational achievement. Some scholars argue that if cognitive enhancement and other neural modifiers could reduce existing inequities, then justice *requires* interventions.[127] This might prove to be so. At the very least, new forms of safe and effective interventions that deliver real advantages to those who use them should not be distributed so as to exacerbate or amplify existing inequities.

Access to low-technology neural modifiers, such as educational enrichment, test preparation courses, or adequate childhood nutrition, raise justice and equity concerns parallel to those raised by novel neural modifiers. Societal tolerance of inequity in access to other crucial goods does not make inequity right, nor should it hamper our efforts to reduce or eliminate inequity where we can.[128] If safe and effective novel forms of cognitive enhancement become available, they will present an opportunity to insist on a distribution that is fair and just. While not eliminating all other less tractable forms of injustice in the distribution of neural health and wellbeing, it is possible to ensure that any new forms of safe and beneficial neural modification do not worsen those injustices.

Moral Agency and Human Dignity

Moral agents are individuals capable of acting freely and making judgments for which they can be praised, blamed, or held responsible. Respect for human dignity has grounded longstanding ethical prohibitions against coerced uses of drugs and devices to alter the brain and nervous system.[129] In addition, some scholars contend that cognitive enhancement and other neural modifications also pose a potential threat to moral agency and human dignity.[130] Enabling individuals without specific impairments to achieve higher levels of cognitive function is ethically controversial. Scholars question whether humans should exercise so much control over the natural world, and debate where to draw the line.[131] On this view, advances that vastly improve human beings cross ethical lines by risking the creation of not better humans but transhumans.

Use of pharmaceuticals to improve alertness, attention, mood, and happiness also raise concerns about morally legitimate paths to success and wellbeing. Some scholars consider achieving success with the help of a pill akin to cheating or taking the easy way out, because they believe success is supposed to be the result of personal effort and hard work. According to this objection, some forms of neural modification offer only false visions of human achievement. This type of success might be valuable for the immediate outcome, but cannot be considered the kind of achievement that results from personal will and exertion. From this perspective, success is as much about *how* goals are achieved as achieving them.[132]

Similarly, some scholars contend that happiness and wellbeing are supposed to be rewards of virtue and good character, not an outcome of medication.[133] Although it can be deeply upsetting and profoundly life-changing to live with traumatic memories, some view medications to dampen memories as problematic because they could prevent individuals from coming to terms with their lives as continuous subjects of both good and bad experiences.[134] From this perspective, neural modification, particularly through pharmacological management, threatens to provide only "fraudulent happiness."[135] Yet, from another perspective, ethical merit might exist in "fraudulent happiness" that enables individuals to be functional parents, providers, and engaged citizens.

In contrast, others view the practice of novel neurotechnologies being used to enhance humans as technological progress and innovation. Scholars evaluate some forms of neural modification for their potential to be used for moral enhancement. Drugs that free us of rage, impulsivity, and aggression might enable us to participate successfully in the moral community. For example, some research results indicate that oxytocin can promote generosity, and other pharmacological substances have been reported to increase cooperation.[136]

Importantly, the empirical evidence supporting the possibility of moral enhancement is thin, and interpreting results in terms of moral enhancement has been criticized by both those internal and external to the scientific community.[137] Some scholars question whether we would be morally better at all if only through use of a drug—our conduct would be the result of a will controlled by the external and artificial stimulus of a pharmaceutical rather than will disciplined through effort. In contrast, others point out that, although technological moral enhancement is only a distant prospect, it can serve as a complement to, not a replacement of, traditional social and educational modes of moral improvement.[138]

Importance of Public Education and Deliberation

In its 2010 report on synthetic biology, *New Directions: The Ethics of Synthetic Biology and Emerging Technologies* (*New Directions*), the Bioethics Commission recognized the importance of informed and reasonable public debate about potentially controversial issues in science, technology, and ethics, emphasizing the importance of an informed public to facilitate democratic deliberation.[139] In this report, it recognizes that the debate about cognitive enhancement and other novel neural modifiers can be fraught with exaggeration and misinformation. Scientists, the media, policymakers, and other stakeholders often hype or inaccurately portray facts.[140] A deeper and more accurate understanding of relevant evidence, realistic potential, and the true and pressing ethical concerns surrounding controversial topics is essential.

In *New Directions*, the Bioethics Commission recommended that a mechanism be created to fact check the diverse claims made about advances in synthetic biology.[141] Similarly, in the cognitive enhancement and novel neural modification arena, a mechanism analogous to FactCheck.org would be a useful tool to facilitate a more informed consumer and public.[142] Educated public debate about science helps to air all relevant perspectives, and participation by the scientific community helps maintain an educated public.[143] This fosters democratic deliberation—collaborative decision making that embraces respectful debate of opposing views and active participation by members of the public. The Bioethics Commission urges our society to uphold this particularly important ethical principle to maximize the potential that emerging neuroscientific discoveries will be well-understood and used for the betterment of the human condition.[144]

Recommendations

Through debates about cognitive enhancement, neural modification has captured the public's imagination. Neural modification includes basic strategies already demonstrated to improve brain and nervous system function. These evidence-based strategies include healthy diet, exercise, sleep, and education. Novel neural modification methods foster hope that we soon will be able to ameliorate the symptoms of depression, curtail addiction, and prevent dementia-induced cognitive decline, among other health benefits. In addition, contemplating novel methods to improve such functions as learning and memory in school or performance in competitive professions is truly exciting.

Public discourse around neural modification, including cognitive enhancement, reflects fascination, but also raises ethical concerns. Although unjust access to beneficial interventions could exacerbate social inequities, some interventions also could promote equity by closing existing gaps. Fully appreciating any risks in addition to potential benefits inherent in using neural modifiers is imperative. Guidelines for practitioners and relevant stakeholders can help guide use of brain and nervous system interventions and their potential risks and benefits in diverse circumstances.

* * *

Several well-known lifestyle interventions, such as adequate sleep, exercise, and nutrition, are associated with improved neural function. Similarly, public health interventions, such as lead paint abatement, can help prevent the negative impact of environmental exposures on neural development and function.[145] These behaviors and conditions can maintain and improve neural health and might be safer and more effective than those offered by novel neural modifiers, some of which might have minimal benefits and uncertain risks.[146]

Recommendation 1: Prioritize Existing Strategies to Maintain and Improve Neural Health

In addition to developing new drugs and devices to maintain and improve neural health, funders should prioritize and support research on existing, low-technology strategies, such as healthy diet, adequate exercise and sleep, lead paint abatement, high-quality educational opportunities, and toxin-free workplaces and housing.

Implementation of evidence-based strategies to maintain or improve cognitive, motor, and nervous system function through behavior or environmental modification should be prioritized over strategies based on novel and often very expensive interventions for which evidence of effectiveness remains uncertain. Funders should allocate resources to study and implement strategies—both low- and high-technology—that support neural health. For example, a long history of studies demonstrates that education and adequate sleep can improve cognitive function.[147] Neuroscience can help us better understand the mechanisms for learning and sleep, the effects that these activities have on the brain, and how to make them more efficient and effective. A thorough, neuroscientific understanding of known interventions can contribute to implementation of effective strategies for improving neural functioning across society.

*

Existing treatments for neurological disorders are valuable and can be improved. In addition, emerging neural modification interventions will help reduce the individual and societal burden of neurological disorders. For example, brain-computer interfaces might allow those paralyzed from spinal cord injury, brainstem stroke, Lou Gehrig's disease, or other disorders to perform complex and flexible movements with a neurally controlled robotic arm.[148] Early research indicates that brain stimulation techniques might alleviate symptoms in wide-ranging neurological disorders and psychiatric conditions, and some FDA-approved brain stimulation devices are currently being used to treat conditions like bipolar disorder, but ethical concerns remain.[149] Safe and effective treatments can improve the lives of millions of individuals living with such conditions.

Recommendation 2: Prioritize Treatment of Neurological Disorders

Funders should prioritize research to treat neurological disorders to improve health and alleviate suffering. This research should consider individual, familial, and public health burdens as well as potential risks, benefits, and long-term effects of specific interventions.

Research on interventions to treat neurological disorders should be prioritized. Funders should allocate resources to study and implement productive, low-technology strategies in addition to developing new treatment interventions.

*

Although the Bioethics Commission recognizes the need to prioritize the study of both traditional and novel interventions for the prevention and treatment of neurologic disorders, it nonetheless also supports research to better characterize and understand novel neural modification techniques to augment or enhance function. Limited, inconclusive evidence exists for the benefits and risks of stimulant drugs, such as Ritalin® and Adderall®, and brain stimulation methods, such as tDCS, as neural enhancers.[150] In addition, few data are available on the prevalence of the use of neural modification interventions for cognitive enhancement purposes. Most prevalence data for stimulant drug use are limited to specific populations, such as college students, and prevalence surveys generally do not capture why individuals use the drugs non-medically (i.e., whether for cognitive enhancement or other reasons).[151] Only limited anecdotal evidence exists on the use of brain stimulation devices for enhancement purposes.[152]

Recommendation 3: Study Novel Neural Modifiers to Augment or Enhance Neural Function

Funders should support research on the prevalence, benefits, and risks of novel neural modifiers to guide the ethical use of interventions to augment or enhance neural function.

The lack or misinterpretation of evidence on prevalence, benefits, and risks can contribute to exaggerated expectations and pronouncements surrounding neural modifiers, making them seem more widespread, effective, or threatening than they are. Ethical analyses must account for limitations of available evidence.[153] Targeted research will provide better evidence to ensure that an accurate message is portrayed to the public about the potential impact of these interventions. This research should consider the prevalence of use in a variety of educational and professional settings, potential risks involved in the use of specific interventions, long-term effects, and effectiveness in real-world settings.

Better evidence is needed on which to base ethical deliberations. For example, concerns about human dignity, distributive justice, and fair access to enhancing interventions will be most salient if and when neural enhancement interventions are demonstrated to be both beneficial and safe in enhancing neural function. Use of neural modification interventions should be supported by robust ethical deliberation guided by ample evidence on the benefits and risks of each intervention.

*

One prominent ethical consideration is access to the benefits of neural modification. Limiting access to effective enhancement interventions to those who already enjoy greater access to other social goods would be unjust. It also might deprive society of other benefits of more widespread enhancement—including societal benefits, such as improved civic engagement or greater productivity—that increase as more individuals have access to the intervention.[154] In addition, more widespread enhancement might help to close some gaps in opportunity that are related to neural function, such as educational attainment or employment. These potential benefits support the claim that access to safe and effective enhancement interventions should not be limited to those with financial or other means.

Recommendation 4: Ensure Equitable Access to Novel Neural Modifiers to Augment or Enhance Neural Function

Policymakers and other stakeholders should ensure that access to beneficial, safe, effective, and morally acceptable novel neural modifiers to augment or enhance neural function is equitable so as not to compound or exacerbate social and economic inequities.

One way to address concerns about justice is to ensure that proven enhancement interventions are available to everyone or to no one; however, both of these extremes can be unjust. Making enhancements available to everyone, although fair, might simply preserve existing inequities. Making enhancements available to no one, although similarly fair, might deprive individuals and society of the potential benefits the intervention could bring. Although limited, evidence indicates that some cognitive enhancement technologies might confer the greatest benefit on those most in need. If this is demonstrated to be the case, these technologies might reduce gaps in cognitive performance that can have substantial implications for an individual's social and economic position.[155]

Evidence about who benefits most from neural enhancements can guide policies that are sensitive to the contours of social and economic disparities.[156] Enhancement interventions that are demonstrated to be safe and effective should be assessed to determine their potential to affect social and economic disparities. Just as additional evidence is needed to understand what interventions—if any—are beneficial, safe, and effective, additional debate is needed to inform whether specific interventions are morally acceptable.

*

Clinicians often receive requests to prescribe medications for cognitive enhancement.[157] Some guidance suggests that prescribing medications to adults for the purposes of cognitive enhancement can be considered ethically permissible, yet individual clinicians must decide whether to prescribe the medication to particular patients.[158] These decisions are more ethically complex with regard to children, because children lack legal and ethical consent capacity and are vulnerable to coercion.[159]

Clinicians considering prescribing interventions for neural modification, including cognitive enhancement, should have access to detailed professional guidelines that can help them manage patient requests ethically, especially with regard to children and adolescents.[160] Other stakeholders also would benefit from education and guidance on neural modification interventions. These stakeholders include employers, parents, educators, and professional organizations in fields such as aviation, medicine, and the military, among others, that are associated with on-the-job use of brain and nervous system enhancement interventions.

Recommendation 5: Create Guidance About the Use of Neural Modifiers

Professional organizations and other expert groups should develop guidance for clinicians, employers, parents, educators, and patients about the use of neural modifiers and their potential risks and benefits. Medical professional organizations should develop guidelines to assist clinicians in responding to requests for prescriptions for interventions to expand or augment neural function. Clinicians should not prescribe medications that have uncertain or unproven benefits and risks to augment neural function in children and adolescents who do not have neurological disorders.

Widely diverse groups and professional organizations can contribute to the development of guidelines on the prescription and use of neural modification interventions. Detailed guidance and educational materials for stakeholders can guide decisions about how to reconcile the potential benefits and risks of using a particular neural modifier under various circumstances—for example, an individual with age-associated memory impairment or a parent seeking medication for a healthy child to improve unimpaired cognitive function.[161]

Health care providers are the gatekeepers of many medications requested for enhancement purposes and, through their professional organizations, are well-placed to develop comprehensive guidance on appropriate prescribing practices.[162] Professional organizations in other relevant fields, such as education, guidance counseling, aviation, and medicine, should work to develop policies on the use of enhancement interventions. These policies can guide health care providers' decision making and help to protect individuals from pressure to use enhancement interventions. Policies should inform stakeholders about the ethical concerns that arise with the prescription and use of neural modifiers, including justice, risk, coercion, and respect for human dignity. Guidelines can clarify misunderstandings and prevent ethical missteps.

Generally, clinicians should not prescribe medications that have uncertain or unproven benefits and risks to augment neural function among children and adolescents who do not have neurological disorders.[163] The American Academy of Neurology endorses the position that cognitive enhancement with prescription drugs is not ethically justifiable among a healthy pediatric population.[164] The pediatric population presents unique ethical concerns: Children lack legal and ethical consent capacity, and clinicians and parents must make decisions that account for children's developing autonomy, their right to an open future, and their vulnerability to coercion and undue pressure.[165] The current state of evidence on novel neuroscience interventions to augment or enhance indicates that, generally, they should not be used among the pediatric population.

* * *

By broadening the discussion of cognitive enhancement to include all forms of neural modification, the Bioethics Commission is expanding the scope of the current debate. Neural modification—to maintain or improve brain health within typical or statistically normal ranges, treat neurological disorders, and expand or augment neural function—raises a set of ethical considerations, including justice concerns, questions about how to reconcile risks and benefits on a case-by-case basis, and concerns about personhood and moral agency. The Bioethics Commission asserts that cognitive enhancement using novel neurotechnologies exists on a spectrum with other neural modifiers. Thus, it recommends that stakeholders focus on a set of priorities for developing and using neural modifiers, including prioritizing safe and well-studied methods for neural enhancement, prioritizing the development of neural modifiers that have the potential to treat disorders, and conducting more research on the prevalence, benefits, and risks of novel neurotechnologies.

SECTION 3. CAPACITY AND THE CONSENT PROCESS

Scientific progress to improve human health requires human participation in clinical research. Contemporary neuroscience research promises to provide important insights into the nature of disease, as well as possible prevention strategies, diagnostic tools, and treatments for a range of increasingly prevalent and often devastating neurological disorders and psychiatric conditions. Much-needed neuroscience research is ongoing and will continue to be conducted on disorders that are often associated with impaired consent capacity. Such research often concerns the very organ responsible for decisions about whether to participate in such research in the first place.

Informed consent, based on the principle of respect for persons, is a foundational tenet of clinical and research ethics.[166] It is a widely accepted ethical, legal, and regulatory requirement for most clinical research and health care interactions.[167] To give informed consent, one must have the capacity to provide consent, known as consent capacity. The underlying abilities that consent capacity comprises are debated, but often are thought to include an ability to understand disclosed information, appreciate its significance, and use the information to reason and make and express a choice.[168] Neuroscientists who conduct research involving human participants commonly work with populations or individuals whose consent capacity might be absent, impaired, fluctuating, or in question.[169] Similarly, clinicians in many settings encounter patients with impairments in consent capacity. Widely diverse disorders and injuries can affect an individual's capacity to understand information, consider the benefits and risks of research participation, or reach an informed decision regarding study participation. Neurological disorders, such as head trauma, stroke, dementia, neurological cancers, and metabolic disorders, affect neurological function and can lead to impaired decision-making capacity. Individuals with psychiatric conditions, including schizophrenia or major depression, and those who use psychoactive medications or addictive substances also might have impaired consent capacity. Of note, not all individuals with these conditions have diminished consent capacity or consistently diminished capacity—some affected individuals are capable of understanding information and providing informed consent some or all of the time. In addition, certain novel neuroscientific research interventions can alter participants' consent capacity. For example, procedures such as deep brain stimulation (DBS) or electroconvulsive therapy might pose risks to cognitive function.[170]

> "We don't want to stereotype people based on a diagnosis...[or] disrespect the autonomy of people who are still able to make their own choices. On the other hand, you don't want to fail to protect vulnerable people who can't understand the decisions before them. That would be taking advantage of them for the greater good."
>
> Dresser, R., Daniel Noyes Kirby Professor of Law, Professor of Ethics in Medicine, Washington University, St. Louis. (2014). Neuroscience Research Across Life Stages. Presentation to the Bioethics Commission, June 9. Retrieved February 2, 2015 from http://bioethics.gov/node/3775.

Neuroscience research is a principal means of promoting progress and benefiting populations affected by neurological disorders and psychiatric conditions, including those associated with impaired consent capacity. Substantial progress in understanding many of these disorders has been possible because of the advances in research, yet much more needs to be done. This research is ethically challenging because it requires participation of individuals with disorders associated with impaired consent capacity, and voluntary, informed consent is an important ethical tenet in the protection of research participants. To reconcile these competing commitments, such research should only proceed with additional ethical safeguards and protections in place. Protections might include robust initial and ongoing assessment of consent capacity; methods to improve informed consent to accommodate participants' needs, including audiovisual means and paced verbal instructions; methods to respect assent and dissent when consent capacity is partial or in question; independent consent monitors; limits on risk; clear parameters and procedures for obtaining the permission of a legally authorized representative (LAR) when a participant lacks consent capacity; research advance directives; and stakeholder engagement. In addition, use of research advance directives (a set of written instructions articulated by an individual to appoint a proxy and to direct their involvement in future research) and attempts to mitigate stigma associated with conditions that lead to impaired consent capacity can provide protection.[171] Clear practices that are well-articulated publicly are needed to protect those with impaired consent capacity while promoting vital neuroscience research.

Grappling with the complex challenges surrounding informed consent and consent capacity requires diverse expertise, including perspectives from neurology, psychiatry, psychology, social work, patient advocacy, and bioethics. In addition, the concept of consent capacity, the causes of its impairment and its potential to be restored through therapeutic intervention are areas that stand to benefit greatly from the fruits of neuroscience research. Neuroscience research could help refine our understanding and assessment of decision-making capacity, including consent capacity, and its underlying neurological correlates.[172] Neuroscience research also has the potential to help us understand what abilities and decision making skills are needed for effective consent with individuals affected by a wide range of conditions, and to inform improvements in managing impaired consent capacity. Thus, an ethical analysis of consent capacity is well-suited for this report on neuroscience and ethics.

Ethical Analysis

Contemporary neuroscience research presents an opportunity to achieve a deeper understanding of brain-related disorders that represent a major public health burden and have a severe impact on caregivers and loved ones (see *Section 1: Background and the Promise of Neuroscience Research*, above). Some of these disorders, however, are associated with impaired, fluctuating, or diminishing decision-making abilities, which can affect the individual's capacity to consent. It is vital, wherever possible, to find ways to ethically and responsibly include individuals with impaired consent capacity in neuroscience research, as well as to use neuroscience to better understand the capacities that enable and impede informed consent.

As with many of today's research protections, concern for participants with impaired consent capacity stemmed initially from revelations of past abuse and mistreatment, such as experiments with institutionalized individuals, and widespread public concern regarding psychosurgery.[173] Against this backdrop, national advisory bodies, researchers, and institutional review boards (IRBs)—committees that review human subjects research—have struggled to both protect against future abuses and not unjustly or unnecessarily exclude potential participants because of their condition or impaired consent capacity or for fear of legal liability. This dual mission—protection and inclusion to ensure that the benefits of research are distributed equitably—shapes many core ethical considerations surrounding capacity, the consent process, and participation in research.

Ensuring Access to the Benefits of Research through Inclusion

Several foundational principles of bioethics—respect for persons, beneficence, and justice and fairness—support the inclusion of participants with impaired consent capacity in neuroscience research, with appropriate protections in place.

Inclusion in research of persons who might have impaired consent capacity reflects respect for persons, which encompasses respect for other forms of agency—in other words, even though individuals might lack autonomy, they might have the ability to express certain preferences or participate in some way in decision-making processes. The principle of respect for autonomy calls on us to respect the abilities of others to reason, come to considered judgments, and make decisions regarding what is best for them.[174] The broader principle of respect for persons encompasses more than just respect for autonomy, recognizing that all people, including those who are not autonomous, deserve respect. Respectful research practices can take many forms, including thoughtful implementation of additional protections for participants who lack consent capacity.[175] Providing potential research participants with information about the purpose, prospect of benefit, and risks related to a research protocol enables individuals to take part in deciding whether to participate. Respect for persons is therefore at the core of concerns about consent capacity in research. The principle requires neuroscience researchers to support autonomy and all forms of agency whenever possible.[176] This includes respect for expressions of agency that, when encountered, reflect meaningful participant values or preferences.[177] Respectful practices also include facilitating measures, such as research advance directives, which help individuals express their wishes for the future.[178] In addition, respecting participants who might have impaired, fluctuating, or diminishing consent capacity means making every effort to avoid two equally troubling mistakes: misidentifying capable individuals as incapable, and misidentifying incapable individuals as having consent capacity.[179]

Research benefits can accrue to affected populations if research participation practices are inclusive. Beneficence calls for efforts to secure the wellbeing of others. Public beneficence confers on society an obligation to advance scientific and technological discovery that can improve public wellbeing.[180] Neuroscience research can potentially improve the prevention, diagnosis, and treatment of disorders that can lead to cognitive impairments; however, failing to support neuroscience research on certain disorders because potential participants might have impaired consent capacity can do a disservice to current and future patients.[181] For example, a 2008 study used surrogate consent to enroll participants with impaired consent capacity to study an investigational treatment that later became the only U.S. Food and Drug Administration (FDA)-approved treatment for acute ischemic stroke. Researchers estimated that, had they not been able to include these participants, the research would have taken four times as long and its validity and impact would have been undermined.[182]

When research participation practices are inclusive, fair distribution of research benefits is made possible. Justice and fairness requires that the benefits of neuroscience research be distributed equitably across society.[183] In the previous stroke research example, failing to include persons with impaired consent capacity might have been a disservice to many patients who suffer strokes, because research conclusions would not have been readily generalizable to groups outside the study population.[184] To address the very conditions that threaten to impair consent capacity, we should do our best as a society to find ethical means wherever possible to include individuals with impaired consent capacity—with appropriate protections in place—in potentially path-breaking research.

Protecting All Research Participants

As the history of research ethics reveals, researchers have sometimes selected those with impaired consent capacity for research participation in part because of their greater convenience as a participant pool.[185] Ethical human subjects research can involve treating people as means as well as ends-in-themselves, whereas, when participants are exploited only to further the interests of others, they are being used as *mere* means.[186] That is ethically unacceptable. Prohibiting the exploitative treatment of others as mere means to scientific advancement is critically important and at the heart of current research oversight. Development of additional ethical protections is meant to help avoid future transgressions.[187]

Subpart A of the U.S. Department of Health and Human Services' (HHS) regulations, Protection of Human Subjects (codified at 45 C.F.R. Part 46), also known as the Common Rule, has been adopted by 18 federal departments and agencies that conduct or fund human subjects research and provides standards for the ethical conduct of federally supported human subjects research. It identifies "mentally disabled persons" as a population that is potentially more vulnerable to coercion or undue influence.[188] Research ethics often invokes the concept of vulnerability to highlight the unique needs of certain populations who participate in research. Importantly, this framework of vulnerability highlights the ethical goal of providing special protections to those participants who might be more susceptible to exploitation or harm than others as a result of research participation. The concept of vulnerability in research ethics is longstanding and complicated.[189]

Approaches to participant protection based on considering members of entire groups vulnerable can have unintended consequences.[190] Labeling those with specific diagnoses as vulnerable can be potentially stigmatizing by reinforcing gross generalizations about large and varied groups of people.[191] Some successful initiatives to combat stigma have focused on frameworks of empowerment, which encourage individuals or their loved ones to advocate for ethical policies and practices.[192] Nevertheless, invoking the concept of vulnerability serves a vital practical and ethical function—it calls our attention to research with human participants that warrants special scrutiny.[193]

Protecting research participants also requires preventing exploitation—taking unfair advantage of another.[194] In research, informed consent helps protect against exploitation by providing potential participants with information about the ramifications of participation.[195] When informed consent cannot serve this function, identifying and establishing alternative or additional protections—especially to prevent clear-cut cases of exploitation—is central to proceeding ethically with research.

The challenge for policymakers and others is to delineate necessary standards for protection while enabling versatile and responsive policies to adapt protections to the needs of research participants in highly variable circumstances.[196] Scholars have referred to the fundamental tension between under- or overprotection as the "pendulum" of human subjects research protections.[197] Ethical and regulatory policies formed in reaction to unethical research practices tend to focus on the risks of research rather than the benefits and lean toward excluding potentially vulnerable participants. Many stakeholders in the field see exclusion as overprotective. Over time, policy has shifted toward maximizing inclusion—a movement that potentially discounts the risks of participation or increases the risk of exploitation.[198]

Neuroscience researchers must walk a fine line between inclusion and protection. Key to walking this line is ensuring that participants are not being exploited or otherwise being used as mere means to an end. Ethical human subjects research can involve consent by participants or permission from legitimate surrogates to serve as a means to achieve broader societal goals—after all, the goal of research is to produce generalizable knowledge, not to provide therapeutic benefit to study participants. However, treating participants as mere means violates respect for persons.[199] The ethical safeguards discussed below illustrate some of the steps neuroscientists and others can take to ensure that their treatment of participants with impaired consent capacity is not exploitative. Fully informed consent by the participants themselves might not be possible in certain cases, but participants can still be respected and allowed to participate with adequate protections in place.

Avoiding and Alleviating Stigma

When considering and addressing ethical concerns about consent capacity, avoiding policies and practices that perpetuate or exacerbate stigma is crucial. Impaired consent capacity is associated with numerous and diverse health conditions. Researchers often make assumptions concerning an individual's abilities on the basis of broad generalizations concerning a health condition, thus equating certain diagnoses with a lack of consent capacity. These assumptions are particularly prevalent in social attitudes toward those with mental illness diagnoses.[200] Such unfounded, often unexamined, beliefs can subject these individuals to stigmatizing and discriminatory practices, in which their individual needs and capacities are not adequately assessed or respected. Stigma can negatively affect an individual's quality of life through a hindered recovery, loss of legal rights, medical care discrimination, and a shorter life span.[201]

Several national advisory bodies have recommended policies to protect potential participants with certain conditions that are associated with impaired consent capacity.[202] However, focusing on groups with certain conditions instead of individuals who might have impaired consent capacity can have unintended, stigmatizing effects. It might imply, for example, that all persons with particular mental illnesses have impaired consent capacity.[203] Although certain conditions might be associated with impaired consent capacity, different individuals with the same diagnosed condition can exhibit varying capacities, depending on environment, relationships, context, severity of disease, and neuropsychological functions. Assumptions about individuals' abilities based on a diagnosis can reflect stereotypes that undermine the respect due to those individuals.[204]

Ethical neuroscience research can help mitigate stigma and discrimination. For example, neuroscience draws attention to multiple neurological disorders that can impair consent capacity, allowing for a more robust discussion that is not limited to mental illness. Scholars note that effective campaigns to reduce stigma and discrimination must be targeted, local, credible, and continuous.[205] Efforts that incorporate or originate from the perspectives and experience of individuals affected by such disorders are especially effective.[206] Neuroscience research on consent capacity itself can identify the underlying neural correlates of cognition and can illuminate how cognitive capacities, including consent capacity, are contingent upon environmental and social cues. By providing a more accurate picture of the diversity and determinants of human abilities, neuroscience might help to undermine common assumptions about what individuals with stigmatized conditions can do and contribute. However, because socially stigmatizing attitudes are not simple to eliminate, education regarding neuroscientific facts alone cannot counter stigma.[207] Factual information works best when coupled with other strategies, such as increased contact with those who live with stigmatized conditions.[208]

History of U.S. Policy Proposals and Recommendations

Understanding, evaluating, and improving informed consent processes is an ongoing goal of research ethics. The history of research ethics includes multiple efforts by national-level advisory bodies to provide guidance for research involving individuals who might have compromised or impaired consent capacity. This section describes historical efforts from the 1970s to the present, and summarizes previous attempts at proposing regulations or guidance about research involving individuals who might lack consent capacity (see *Appendix I: History of Major U.S. Policy Proposals and Recommendations on Consent Capacity in Research* for a timeline of the events described here, and relevant others). This history illustrates the challenging tension between the need for rigorous research on diseases and disorders with high morbidity and mortality, and the need to protect individuals who might be vulnerable because of impaired consent capacity. Federal regulations specific to research involving individuals with impaired consent capacity have never been adopted.

In 1977, the National Commission for the Protection of Human Subjects of Biomedical and Behavioral Research (National Commission) released a report entitled *Psychosurgery*, which outlined ethical protections and limitations on psychosurgery research involving capable individuals as well as those with impaired consent capacity.[209] The following year, the National Commission released *Research Involving Those Institutionalized as Mentally Infirm*. That report discussed the need for additional safeguards to protect vulnerable research participants with impaired consent capacity, while recognizing that "prohibiting such research might harm the class of mentally infirm persons as a whole by depriving them of benefits they could have received if the research had proceeded."[210] These reports constituted a response to public revelations about unethical research that had occurred in the preceding decades, often with groups who were institutionalized or who could not provide valid informed consent.

In response to these two reports, the U.S. Department of Health, Education, and Welfare (which later became HHS) drafted proposed regulations to guide research involving adults who lack consent capacity, consistent with the National Commission's recommendations. Although the reason is unclear, those draft regulations were not adopted.[211] Resistance to those recommendations might have stemmed, in part, from their potentially overprotective nature, which could stall potentially valuable and ethical research.

In the late 1990s, the National Bioethics Advisory Commission (NBAC) revisited the topic. In *Research Involving Persons with Mental Disorders That May Affect Decisionmaking Capacity*, NBAC made 21 recommendations in six categories: review bodies; research design; informed consent and capacity; categories of research; surrogate decision making; and education, research, and support—several of which included proposals for new regulations.[212] NBAC's recommendations did not result in new HHS policies or regulations.[213] Scholars have speculated about this lack of regulatory uptake. Although NBAC recognized that mental illness should not be equated with impaired consent capacity, some scholars have criticized its narrow focus on diminished capacity among individuals with mental disorders.[214] Critics have noted that NBAC's limited focus could stigmatize individuals with mental disorders by implying they are likely to exhibit impaired consent capacity—a concern that NBAC acknowledged in its report. Critics supported a wider focus to ensure the protection of all research participants who might have impaired decision making, regardless of diagnosis.[215] In addition, NBAC recognized the importance of diversity among IRB members, but scholars have raised concerns about the lack of input from a diversity of researchers and patient groups in NBAC's own deliberations.[216]

Some state legislatures introduced bills modeled after NBAC's recommendations, but ultimately none were enacted into law.[217] However, in 1999, the National Institutes of Health (NIH) released interim guidance for IRBs that was "generally consistent with the NBAC report."[218] NIH's guidance, updated in 2009, provides researchers and IRBs with points to consider when conducting research involving individuals who might have impaired consent capacity.[219] This guidance is intended to help researchers understand how to comply with federal and state regulations and to facilitate consideration of ethical concerns and the dual mission of inclusion and protection.

In 2009, the Secretary's Advisory Committee on Human Research Protections (SACHRP) addressed research involving individuals who might have impaired consent capacity.[220] SACHRP provided 10 recommendations on several topics, including consent capacity, IRB membership and procedures, participant selection, and LARs. In addition, it proposed a regulatory solution for defining who can serve as an LAR.[221] This guidance reached beyond mental health or psychiatric conditions to address the wide array of conditions that can lead to impairments in consent capacity.

Similar to the proposals that preceded it, SACHRP's recommendations were not incorporated as official guidance or regulations for researchers. Yet SACHRP's recommendations are widely cited in current discourse on appropriate additional protections. SACHRP's delineation of who can serve as an LAR for research is often highlighted, given the lack of relevant legislation in most states.[222]

Current Regulatory Framework

No federal regulations directly address research participation of adults with impaired consent capacity. The Common Rule requires oversight of research by IRBs and requires voluntary informed consent from participants or permission from their LARs.[223] The Common Rule also requires "additional safeguards" when participants might be vulnerable for various reasons, including mental disability, but does not stipulate what these safeguards should be.[224] Protections for certain vulnerable populations exist: Subpart B describes protections for pregnant women and fetuses; Subpart C describes protections for prisoners; and Subpart D describes protections for children.[225] Although previous advisory bodies recommended regulatory changes related to research involving adults with impaired consent capacity, no specific regulations have been promulgated.

> "[R]ight now…comparing it to children or prisoners or fetuses, we actually don't have subparts in our regulations even addressing what the rules should be for decisionally incapacitated adults."

Menikoff, J., Director, Office for Human Research Protections, U.S. Department of Health and Human Services. (2014). Neuroscience Research and Diminished Capacity. Presentation to the Bioethics Commission, August 20. Retrieved February 3, 2015 from http://bioethics.gov/node/4005.

Although no specific HHS regulations exist for research involving individuals with impaired consent capacity, as described above, regulations require additional safeguards for vulnerable populations, including "mentally disabled persons."[226] The HHS Office for Human Research Protections (OHRP) offers some guidance and clarification of how the federal regulations apply to research involving individuals with impaired consent capacity.[227] OHRP notes that the HHS regulations allow an LAR to enroll individuals who cannot provide their own consent into research protocols.[228] In addition, in guidance about research involving participants with potentially impaired consent capacity, OHRP emphasizes that federal regulations require that IRBs possess the necessary professional competence to review research activities, either through IRB members with appropriate experience and expertise or invited consultants.[229] Similar requirements exist in FDA regulations on protection of human subjects.[230] In 2014, FDA released draft guidance on informed consent for research, in which one section addresses research involving participants with impaired consent capacity.[231]

The FDA draft guidance (similar to the NIH guidance document described above) leaves the decision about including individuals who might lack consent capacity to the discretion of IRBs and investigators, and provides several considerations to help address challenges that might arise when enrolling such participants in clinical studies.[232]

A patchwork of applicable legal protections also exists at the state level. Current laws for designating an LAR to facilitate decisions about medical or clinical care vary by state.[233] For example, they differ in describing how LARs should make decisions on behalf of patients in the clinical context and who can serve as an LAR. Very few of these state laws address decisions about enrollment in research.[234]

Within this legal and regulatory framework, uncertainty and lack of clarity remain regarding ethically acceptable research involving participants with potentially impaired consent capacity. A survey of U.S. IRBs revealed considerable variability in IRB policies and practices.[235] Although some argue that the flexibility and discretion granted to IRBs and researchers by federal regulations are essential for the range of valuable research that can be done, others note that uncertainty about how to protect individuals with impaired consent capacity results in inconsistent practices that are either too restrictive or too permissive.[236]

Additional Ethical Safeguards

The Common Rule, FDA regulations and guidance, NIH guidance, and many institutional-level policies and guidance call for additional safeguards for vulnerable research participants, including adults with impaired consent capacity. Respectful and just research policies and practices demand both fair inclusion and additional safeguards for prospective research participants with impaired consent capacity. Relevant safeguards might include assessment of consent capacity, solicitation of assent and respecting dissent, use of independent monitors, potential limits on allowable risk, processes to designate and seek permission of an LAR, research advance directives, and stakeholder engagement.

Consent Capacity Assessment and Modified Procedures

Respectful policies and practices acknowledge the need to consider consent capacity individually, among diverse participants, and not make blanket capacity determinations applied to all persons with specific diagnoses. Consent capacity entails the ability to understand pertinent information, appreciate its significance, and use the information to reason and make and express a choice. Consent capacity exists along a continuum and varies among individuals with widely diverse disorders. Investigators should assess consent capacity to avoid making assumptions about prospective participants' consent capacities on the basis of a diagnosed disorder and thus avoid unfairly labeling and stigmatizing individuals and groups. Robust capacity assessment before research begins (and when indicated, during research) also helps ensure that participants with impaired, fluctuating, or diminishing consent capacity are adequately protected.

Investigators conducting research involving individuals who might have impaired consent capacity should ascertain if potential participants have this capacity. For most clinical research, investigators assess consent capacity informally.[237] Validated assessment tools are available for assessing decisional capacity in both the research and treatment settings. Advantages of these tools include the ability to formalize an otherwise intuitive process at a low cost and to do so relatively quickly.[238] However, the tools vary in content and scoring and have variable validity and reliability.[239] An example of an established tool for assessing consent capacity in research is the MacArthur Competency Assessment Tool for Clinical Research (MacCAT-CR).[240] Researchers can tailor the questions in this instrument to a specific research protocol and obtain a score indicating whether the individual has the capacity to consent.

Regardless of which assessment tool is used, researchers and IRBs should consider whether the participants' condition, intervention under study, or other environmental factors indicate a need for assessment and reassessment of consent capacity during the course of research.[241] Consent capacity is task-specific and depends on the nature and complexity of the decision at hand. An individual might have consent capacity for certain studies or procedures but not for others. Consent capacity also can fluctuate and might improve or worsen as the individual's condition changes. For example, with regard to the task-specific standard for capacity, the decision-making capacity of an individual with dementia might be sufficient to meet the ethical and legal standard for a clinical intervention that is therapeutic, but not for an experimental protocol that promises no direct therapeutic benefit. Similarly, a potential participant might have the capacity to consent to a simple and easily understood protocol, but might lack capacity to consent to a more complex protocol. In addition, decisional capacity can fluctuate, improving or worsening over time, or it can diminish over time as disease progresses, as is often the case with dementia.

An individual's understanding of the information needed to make a decision depends in part on how the information is presented and explained. Modifying informed consent processes by simplifying forms, orally explaining study procedures, or using creative strategies, such as multimedia supplements, might improve comprehension among participants with certain cognitive or decisional impairments.[242] Understanding more about the underlying causes of impaired decision making could lead to development and testing of effective consent and assessment strategies, such as corrective feedback, repeated explanation, or multimedia consent techniques.[243] Finding effective strategies also could be useful for assessment and consent in the clinical environment.

"[W]henever possible...when we see people who have some degree of decisional impairment, rather than jumping to the conclusion that they can't make decisions for themselves, we ought first to try to ameliorate their impairments so as to enable them to make those decisions before we turn those decisions over to other people."

Appelbaum, P.S., Elizabeth K. Dollard Professor of Psychiatry, Medicine & Law; Director, Division of Psychiatry, Law, and Ethics; Director, Center for Research on Ethical, Legal & Social Implications of Psychiatric, Neurologic & Behavioral Genetics, Department of Psychiatry, Columbia University. (2014). Assessing Decisional Capacity in Neuroscience Research. Presentation to the Bioethics Commission, August 20. Retrieved February 3, 2015 from http://bioethics.gov/node/4005.

Assent and Dissent

When LARs make research decisions on behalf of individuals who lack consent capacity, researchers should include the individuals in the informed consent process to the extent possible. Many participants lacking consent capacity can still express meaningful desires regarding research procedures, including by indicating assent or dissent. For example, some scholars have proposed seeking participants' assent and respecting dissent to conduct dementia research in accordance with ethical principles.[244] Respecting dissent serves as a protective measure to avoid inflicting burdens and maintains the dignity of all persons in research.[245] Importantly, seeking assent is not the ethical equivalent of seeking informed consent, because it cannot tell us whether risks to participants are being voluntarily accepted on behalf of others. Nevertheless, meaningful expressions of assent and dissent are salient, even if insufficient, evidence of participants' perspectives regarding decisions made on their behalf.[246]

"[I]n the context of the research that I do, you need to have a higher bar for saying yes than you do for no, so [research participants] need to definitely have some capacity to be able to say yes to the research, but [with] even minimal capacity they can say no."

Wright, D.W., Associate Professor of Emergency Medicine, Director of Emergency Neurosciences, Emory University School of Medicine. (2014). Roundtable Discussion. Presentation to the Bioethics Commission, June 9. Retrieved February 3, 2015 from http://bioethics.gov/node/3780.

Uncertainty regarding the assent and dissent of adults who lack consent capacity persists among both researchers and regulators. The ethical significance of assent and dissent are not equivalent. Assent alone does not provide justification to proceed with research. Although participants can meaningfully indicate assent through a wide range of behaviors, to protect participants from exploitation, passive lack of objection should not be interpreted as assent.[247] Meanwhile, dissent can be expressed as a verbal or nonverbal indication of unwillingness to participate in study procedures and gives researchers a strong reason to refrain from proceeding with research. As an additional form of protection, respect for dissent sets a low threshold for tolerating expressions of discomfort by participants at any time during the course of a study.[248] Uncertainty about the standard of meaningful assent indicates the need for further ethical inquiry, which will help articulate and defend conceptual and ethical standards of meaningful assent and dissent. Such research dovetails with other neuroscience research, which might help delineate evidentiary standards that reflect how we know that certain behaviors correlate to meaningful attitudes.

Independent Monitors

Scholars have argued for and against the necessity and desirability of employing independent monitors for research protocols to assess regulatory compliance and such research ethics issues as informed consent.[249] In the context of neuroscience research that might involve participants with impaired consent capacity, or whose capacity is in question, independent monitors can be useful to help researchers consider and address challenges that arise. For example, in some cases, researchers might assume that all participants with a particular disorder lack consent capacity. Independent monitors can facilitate researcher respect for potential participants who might have consent capacity and can exercise their autonomy. In addition, some scholars have suggested that independent monitors can help prevent coercion on the part of researchers who might consciously or subconsciously be motivated by conflicts of interest.[250] Independent monitors who are trained to recognize complexity and nuance can help researchers maintain the highest ethical standards.

Federal regulations authorize IRBs to observe or have an independent third party observe the consent process, including assessment of consent capacity and assent or dissent throughout the research protocol.[251] An independent third-party monitor for consent capacity assessments might be valuable in protocols that have the potential to incur serious risk to participants.[252] Monitors can be trained to observe both verbal and nonverbal cues.[253] They can monitor assent and dissent of participants with impaired consent capacity throughout research and can help determine whether to halt the research with particular individuals on the basis of distress or dissent. Some authors report that independent consent monitors can be a central part of the additional protections in place for certain kinds of research including participants with impaired consent capacity.[254]

Limits on Acceptable Levels of Risk

Respect for persons recognizes persons as autonomous and capable of deliberating about personal goals, considering choices and opinions, and determining their own lives. Respect for persons also establishes that "persons with diminished autonomy are entitled to protection," although some scholars contend that this protection actually stems from the principle of beneficence.[255] In all research, risk to participants must be minimized. Scholars argue that, for research involving participants with impaired consent capacity, research should only move forward if it presents risk below a certain ceiling.[256] The fully informed, autonomous participation of adults in research that poses risk but no prospect of direct benefit furthers important research goals. However, because they have made their own determination about whether participating comports with their values, they are not a *mere* means. By contrast, potential research participants who lack consent capacity are impaired in their ability to make judgments based on their own values and are at risk of being used as mere means. Limiting risk level for participants with impaired consent capacity can help protect participants from exploitation, by seeking to avoid the most obvious forms of exploitation in which social benefits that result from research are gained regardless of the expense to participants. Limits on risk help prevent some individuals from unknowingly bearing egregious risks of harm for the benefit of others.

The diversity of valuable research involving individuals with potentially impaired consent capacity poses a challenge to uniform recommendations for protecting these participants by limiting risk; research protocols can vary widely in the nature and degree of risk participants might confront. Several advisory bodies have recommended limitations on the level of risk to which adults with impaired consent capacity can be exposed in a particular protocol.[257] Some experts have proposed limiting participation to protocols with the potential for direct therapeutic benefit.[258] Such safeguards are similar to protections in place for research involving other vulnerable groups (such as children) who cannot protect their own interests through informed consent. In the case of research involving children, regulations generally only allow research to proceed if it poses no more than minimal risk or offers a possibility of direct benefit to participants; otherwise, regulations require more stringent safeguards.[259]

Some advisory groups recommend that participants with impaired consent capacity not be included in research without the prospect of direct benefit, unless that research is of "vital importance."[260] Other groups propose that researchers only recruit participants with impaired consent capacity if the research is relevant to their disorder.[261] Still other advisory bodies have supported the inclusion of participants with impaired consent capacity in minimal risk research, recognizing exclusion as disrespectful and possibly unjust.[262] Less agreement exists regarding the conditions under which it might be acceptable to enroll individuals with impaired consent capacity in research that poses greater than minimal risk and does not offer a prospect of direct benefit. This determination remains at the discretion of individual IRBs.

Legally Authorized Representatives

Participants with impaired consent capacity can be enrolled in certain kinds of research by an LAR. LARs, sometimes referred to as surrogate or proxy decision makers, are individuals with the legal power to make decisions on behalf of others. State laws dictate who can serve as an LAR, how much decision-making power an LAR has, what kinds of decisions the LAR can make, and what processes and procedures are required to establish an LAR. Using an LAR is one important way to facilitate inclusion of participants with impaired consent capacity in research, ensuring the just distribution of the benefits that might accrue to people who share the disorder under study. Using an LAR also is a reasonable way to help protect participants from exploitation, because loved ones or caregivers who have been designated as LARs (as discussed below) are often the best proxy for representing participant interests.

State laws vary regarding who can serve as an LAR. In most states, health care proxies, or those holding a durable power of attorney for health care previously appointed by individuals when they were capable, are deemed the most appropriate LARs. State laws usually include a list of possible LARs in a hierarchy, including those with health care power of attorney followed by the individual's next of kin (e.g., spouse, adult children, parents, and siblings). Most state laws describe LARs as having authority for medical decision making, but do not indicate whether the LAR's decision-making power applies to research participation. Although OHRP guidance indicates that LARs appointed for medical care can make certain research enrollment decisions under applicable state law, uncertainty remains about whether laws specific to medical decisions can or should extend to research decisions. Medical decisions are presumed in most cases to be compatible with the best medical interests of the individual, whereas research enrollment entails procedures or interventions done for reasons other than the individual's medical interests.

LAR decision making is complex. LARs are often confronted with questions about how to make decisions on behalf of their loved ones. For example, how should they best honor the potential participant's prior wishes and enduring interests? Should they do what the potential participant would have wanted before their impairment? Or should they attempt to determine what they would want in the present (and how can this be ascertained)? The dilemma is deep: How does even someone as close as a loving spouse or parent ascertain, and act in accordance with, the values, authenticity, and sense of self of their loved ones who no longer have the capacity to consent for themselves?

Current practice and literature encourage LARs to make decisions based on a "substituted judgment" standard. Under this standard, out of respect for the now impaired persons, LARs make decisions based on what the persons themselves would have chosen.[263] This standard encourages LARs to make decisions for other individuals in accordance with those individuals' preexisting known or presumed values and wishes that are projected forward in time to circumstances under which the individuals no longer have the capacity to consent.[264] However, this is only possible when the individuals' prior values and wishes are known to some extent. Planning for LAR decision making for clinical care is rare and is even rarer for research decisions.

In making decisions for the impaired individual, LARs consider a series of questions to guide their choices, such as, what did the person value in the past? What do they currently value? Did they make known their wishes about medical treatment or research? How can those values be realized in this context?[265] Vague expressions of wishes regarding research participation can be difficult to translate to the unique circumstances and research protocols that might arise.[266] This decision-making process can be even more complex when it involves individuals who have never had the capacity to form or express values or preferences, or determine what might be their authentic self. When the individual's earlier values and preferences are unknown or were not developed or articulated in a way that guides a particular decision, LARs often use the "best interests" standard. This standard is based on the principle of beneficence. Under these circumstances, LARs make decisions that are consistent with the individual's overall best interests.[267]

Evaluation of the research enrollment decision is an additional safeguard when an LAR is making the enrollment decision on behalf of a potential participant. Assessment might include what the LAR understands about the decision they are making, the benefits and risks of the proposed research protocol, motives for enrolling the individual, and their understanding of the values and preferences of the individual for whom they are making decisions. Researchers must remain aware of the possibility of tension between the interests of the LAR and the impaired individual.

Research Advance Directives

In some cases, individuals can prepare an advance directive that specifies their willingness to participate in certain kinds of research before their consent capacity becomes impaired, to be consulted and honored by an LAR. Generally, an advance directive is the designation of a proxy decision maker and a set of written instructions articulated by an individual to direct the actions of others in the future, in case the individual becomes unable to make his or her own decisions. Honoring an individual's preferences as delineated on an advance directive demonstrates respect for that individual—and it facilitates inclusion of participants with impaired consent capacity while also avoiding exploitation by respecting their preexisting wishes that were intentionally projected into the future for the explicit purpose of consenting (or withholding consent) to health care and research.

One type of advance directive included in the laws of all 50 U.S. states is appointment of a power of attorney for health care, sometimes referred to as a health care agent. Research advance directives, although uncommon, would be especially helpful as part of the informed consent process for research in which the prospective participants' consent capacity might predictably become impaired at a later date. For example, the NIH Clinical Center's advance directive for both health care and medical research provides individuals with an opportunity to select broad categories of research in which they would be willing to participate; delineate values, goals, and limitations that should guide their participation in research; and designate a power of attorney to make research decisions.[268]

However, just as in clinical decision making using an advance directive, practical challenges and ethical concerns associated with research advance directives remain. For example, how closely should they be honored when the wishes of potential participants seem to conflict with the wishes they expressed on paper? Which self should take priority: the person who drafted the directive, or the person with a present-day impairment? A clear-cut way of addressing this challenge might not exist; however, when participants' current wishes comport with those of their advance directives, the advance directives perform a crucial ethical and practical purpose. They lay as solid a groundwork as possible for respect for persons with currently impaired consent capacities.

The dilemma is greatest when the current wishes of a potential participant conflict with those expressed in an advance directive. But even under these circumstances, we can recognize the practically and ethically difficult question of either to (1) enroll resistant participants in a research protocol against their current will, or (2) enroll willing participants against their advance directive that was carried out under circumstances deemed ethically and legally legitimate for that very purpose.

Stakeholder Engagement

Stakeholder and community engagement can help improve informed consent processes, build relationships and trust, and increase the likelihood that research findings are relevant for affected communities.[269] Community engagement is particularly important for research that involves underrepresented and potentially stigmatized groups. *Gray Matters*, Vol. 1 discussed stakeholder engagement as a principal model of ethics integration in neuroscience research.[270] Seeking out the perspectives of persons and groups likely to be involved in research or affected by its results offers the potential to bridge different expectations and understandings of neuroscience research.

Many individuals and groups have a stake in research design, implementation, and results. Increasingly, standard practice in different research areas—especially those with contentious past and present social, political, and ethical implications—is to employ various techniques to identify stakeholders, as well as incorporate and address their perspectives and concerns.[271]

Many approaches to engaging the stakeholders affected by neuroscience research exist, including explicit attention to IRB composition, formal advisory groups, participatory research methods, large public or community meetings, and empirical research designed to elicit stakeholder perspectives.[272] NBAC recommended that IRBs include members or consultants who could contribute understanding of the experiences of those with impaired consent capacity, including current or former patients, family members, patient advocates, or experts on a specific patient population or LAR decision making.[273] Funding agencies, such as NIH, encourage forms of stakeholder engagement in research beyond those that pertain to IRB composition.[274]

Gaps in Our Understanding of Consent Capacity and Additional Protections

The burden of neurological disorders is high and expected to increase considerably as the population ages.[275] Thus, research that might involve participants with impaired consent capacity likely will continue to increase, as will clinical encounters with patients who have impaired consent capacity. To conduct such research ethically, it is imperative that we learn as much as possible about impairments in consent capacity and how to improve consent processes to protect participants. Our knowledge about conceptual, empirical, and practical matters relevant to consent capacity and what can be done to establish the most inclusive and protective practices is incomplete. Advances in neuroscience can also help advance understanding of cognitive capacity, decision making, and consent. We illustrate some examples of the gaps in our knowledge below, but emphasize that these examples are not exhaustive of the potential for future research and analysis.

Conceptual Challenge: Defining Consent Capacity and Vulnerability

Although consent capacity is generally understood to encompass multiple factors, including the ability to understand information, appreciate its significance, use information to reason, and make and express a choice about participation, advances in neuroscience reveal that consent capacity should also account for other aspects that might influence decisions. For example, advances in neuroscience research have prompted scholars to consider whether other aspects of consent capacity such as emotion might have been overlooked. Some researchers report that consent capacity is incompletely understood without examining the emotional aspects of decision making, for example, how mood disorders can affect individuals' appreciation of risk, including risk aversion and risk tolerance.[276] Considering emotional aspects of consent capacity might be especially pertinent to neuroscientific interventions known to influence individuals' mood, such as DBS.[277]

Scholars sometimes conflate vulnerability caused by impaired consent capacity with other constraints on decision making, such as desperation resulting from lack of treatment options.[278] Although desperation is distinct from impaired consent capacity, it can affect individuals' decisions about research participation by altering their risk perception. Other evidence indicates that concerns about coercion, desperation, or participants' expectations that research participation confers medical benefit—referred to as the therapeutic misconception—are distinct from whether a person cannot consent because of impaired consent capacity.[279] Continued research to understand the nature of vulnerability and consent capacity could result in better protections for research participants.

Empirical Challenge: Assessment

Comparing consent capacity assessment tools is challenging. Such tools have different definitions of reasoning, are tailored to specific protocols, and require various skills and training to administer.[280] Instruments are evaluated for reliability, including both consistency across users and with the same user over time. Empirical data exist on instruments' validity, sensitivity (ability to detect those with impairments), and specificity (ability to identify those with consent capacity).[281] However, challenges remain in defining consent capacity, enumerating specific skills needed for decision making, and tools for accurately measuring such skills.

Ongoing empirical and conceptual research helps to refine available assessment instruments and is needed to facilitate developing neuroscience. The empirical "gold standard" against which to validate research consent capacity instruments is a psychiatric exam.[282] However, variability exists in psychiatric assessments of capacity.[283] Conceptual challenges also exist. For example, scholars have divergent views about what combination of abilities consent capacity comprises, and identifying required abilities is integral to determining the validity of assessment tools.[284] Researchers must know what abilities to assess to develop good assessment tools. Interdisciplinary expertise is necessary for developing tools that can reliably and accurately measure human abilities in context, as well as determine which score thresholds indicate whether an individual is capable of providing informed consent.[285] These determinations require a variety of expertise, including understanding of the legal and ethical meaning and conditions of consent along with the neurology, psychology, interpersonal and cultural interactions, accommodation, and clinical understanding of the well-specified meaning and conditions of consent. Determining how and when to reassess consent capacity to affirm continued participation requires similar collaboration across empirical and normative disciplines.

Practical Challenge: Developing Standards for Valid Consent and Community Engagement

Some aspects of protections are insufficiently understood because they take place against a backdrop of more general obstacles to ensuring informed consent. For example, even among participants who are capable of providing consent, the therapeutic misconception is thought to be pervasive.[286] Participants also might be subject to a cognitive bias called unrealistic optimism.[287] Researchers are seeking ways to improve information sharing during the consent process and to avoid misunderstandings. Innovative approaches to enhance understanding include shorter and simpler consent forms, and participant engagement to identify what information is best understood, and in what format.[288] Benchmarks for success need to be considered in light of what works in informed consent practices generally.[289]

Similarly unclear is what standards are ethically required for community engagement. Not all neuroscientists will be able to incorporate extensive involvement of community members, for example, because of resource constraints. In other cases, a lack of familiarity with local community leaders and resources might make community engagement difficult. Moreover, even when such engagement is possible, determining what constitutes success is difficult. Such practical challenges have conceptual aspects, including how to determine which individuals or groups constitute a community or who can legitimately represent a community's views. For example, for certain research that requires community consultation, if or how the research should proceed if a substantial or minority portion of a community express reservations is unclear.[290] For researchers and those responsible for research oversight to know more about which additional protections are ethically necessary or sufficient, researchers must consider both the ethical rationale for certain protections and evaluate what protections they provide in practice.

Recommendations

Consent capacity raises complex ethical questions for neuroscience research. For example, recruitment and retention processes must accommodate variations in consent capacity, and the complexity of enrolling individuals with impaired consent capacity makes participant selection especially challenging. Some research results might have implications for members of groups whose historical association with impaired consent capacity has subjected them to stigmatization and discrimination or exclusion from research participation.

* * *

Neuroscientists often conduct research involving participants whose consent capacity might be absent, impaired, fluctuating, or in question, in part because many of the disorders that neuroscience research addresses can affect consent capacity.[291] To advance research that seeks to ameliorate these disorders, researchers will need to include affected individuals in studies with ethical safeguards in place. Participants will include affected individuals at any stage of life, from child participants to adults with traumatic brain injury to older participants with dementia.

Recommendation 6: Responsibly Include Participants with Impaired Consent Capacity in Neuroscience Research

Researchers should responsibly include individuals with impaired consent capacity who stand to benefit from neuroscience research. Participation, with ethical safeguards in place, can ensure progress aimed at understanding and ameliorating neurological disorders and psychiatric conditions.

Ethical and regulatory protections for research participants, which are influenced by historical revelations of unethical research practices, might lead to overprotection of research participants and exclusion of participants who might lack consent capacity. Contemporary neuroscience offers the potential to better understand—and the hope to one day ameliorate and prevent—devastating neurological disorders and psychiatric conditions. However, to realize this potential and fulfill this hope, affected individuals, including those who might have impaired, fluctuating, or diminishing consent capacity, will need to be included in ethical research with adequate protections in place.[292] Responsible inclusion entails compliance with existing regulations, and the use of appropriate additional safeguards, which can vary, depending on the nature of the research and the population being studied. For example, a protocol studying the progression of Alzheimer's disease might involve continued capacity assessment or an independent monitor who evaluates ongoing consent and capacity assessment processes. An IRB overseeing research involving participants with impaired consent capacity might choose to limit the acceptable level of risk imposed by the protocol below a certain ceiling.

The pendulum should not now swing in the extreme other direction (as too often happens in history) of under-protection. Rather, we can and should strongly strive toward responsible inclusion. Public beneficence, justice, and respect for persons can ground neuroscientists' obligation to be as inclusive as possible, consistent with protecting all research participants when designing research protocols. More inclusive practices will ensure that the fruits of neuroscience research reach all individuals who stand to benefit as long as ethical safeguards for all research subjects are squarely in place. In its 2009 report, SACHRP similarly recognized the importance of inclusion to advance scientific discovery and potentially ameliorate suffering.[293]

*

Researchers have made substantial progress in the past decade characterizing and understanding consent capacity. However, gaps remain, and further research can support development of best practices for ethical research involving participants with impaired consent capacity. The Bioethics Commission encourages researchers to initiate studies to fill gaps in knowledge and help develop sound policies and procedures for research involving participants with impaired consent capacity.

Recommendation 7: Support Research on Consent Capacity and Ethical Protections

Funders should support research to address knowledge gaps about impaired consent capacity, including the concept of capacity, brain function and decision-making capacity, current policies and practices, and assessment tools.

Conceptual research to address gaps in our knowledge, including the influence of vulnerability, desperation, and affective states on decision making, could lead to better protections for all research participants. Moreover, empirical research evaluating assessment tools and additional protections for participants with impaired consent capacity can determine whether they are adequately protective. Researchers and oversight bodies should develop and evaluate innovative protections for participants with impaired consent capacity. These protections might include novel ways to improve participant comprehension and creative research designs that tailor informed consent processes based on information gathered during recruitment, all carefully guided by clear and explicit understandings of the conceptual, ethical, and legal meanings of consent.

Results of such studies should be disseminated widely. For example, professional societies like the International Neuroethics Society or the Society for Neuroscience might encourage members to share ethical strategies at meetings. Investigators also can consider reporting in publications the participant protections that they employed. With the assistance of journal editors, publishing protective consent methods separately or within the manuscript itself—for example, as an "ethics methods" section—could help create a collaborative environment in which neuroscientists facilitate and further best practices in research.[294]

*

Equating certain conditions with impaired consent capacity or making unfounded assumptions about individual abilities based on diagnoses can exacerbate or perpetuate stigma. In addition to avoiding such pitfalls, ethical neuroscience research also can foster a more accurate understanding of neurological disorders and mental illnesses and potentially mitigate stigma. One principal approach to help neuroscience researchers alleviate stigma is stakeholder engagement.

Recommendation 8: Engage Stakeholders to Address Stigma Associated with Impaired Consent Capacity

Funders and researchers should engage stakeholders, including members of affected communities, to build understanding of consent capacity and associated diagnoses to mitigate the potential for stigma and discrimination.

Stakeholder engagement is critical to research design, including identifying what to study and assessing how results might be received. It is also an important safeguard to mitigate potential social harms associated with research participation, cultivate trust, and develop mechanisms to address harms that cannot be anticipated. Stakeholders include those with, or at risk for, impaired consent capacity, caregivers, researchers, and community members affected by research. Stakeholders' contributions can help mitigate stigma and discrimination by providing information about the lived experience of those affected by a particular condition. They can reveal the diversity of ways in which a condition might manifest and help dispel common assumptions about certain conditions.

Ethical research involving participants with potentially impaired consent capacity requires that investigators acknowledge the diversity of individual needs, abilities, and relationships with caregivers. Stakeholder engagement provides an additional layer of ethical protection for participants. Researchers gain valuable information about participant or surrogate concerns by engaging stakeholders directly. Stakeholder engagement can help neuroscientists identify and develop practices tailored to specific protocols, disease communities, or categories of impairment. Stakeholder engagement can also guide development of standards for enrolling participants with impaired consent capacity, and it can help determine and address pertinent research questions. During the past two decades, researchers have worked to understand stakeholder perspectives about research participation of persons with impaired consent capacity. Engaging stakeholder communities will help neuroscience researchers uphold ethical standards and craft best practices, remain accountable to the communities with which they work, and foster thoughtful consideration about the potential for stigma and discrimination.

*

Including affected individuals (those with impaired consent capacity and others) in research is vital to fulfill the promise of neuroscience to ameliorate neurological disorders and psychiatric conditions. The Common Rule requires informed consent from research participants or permission from LARs before research can proceed.[295] Thus, an important step in conducting ethical research involving individuals with impaired consent capacity is determining who can serve as an LAR. Federal and state laws lack clarity about how to make such a determination.

OHRP asserts that, in the absence of a state law authorizing who should serve as an LAR for research, state laws that authorize representatives for *medical* decision making "may be relevant if the research involves those medical procedures or medical treatment."[296] However, this nonbinding guidance can leave researchers and IRBs uncertain about who can serve as an LAR.[297] Clarity in identifying LARs will help researchers and IRBs remain accountable to a clear set of ethical and legal standards for enrolling participants with impaired consent capacity in research.

Recommendation 9: Establish Clear Requirements for Identifying Legally Authorized Representatives for Research Participation

State legislatures and federal regulatory bodies should establish clear requirements to identify who can serve as legally authorized representatives for individuals with impaired consent capacity to support their responsible inclusion in research.

Federal bodies can play a role in clarifying how to identify LARs. SACHRP, for example, recommended a list of persons (in order of priority) who can serve as LARs, to be relied upon in the absence of applicable state law.[298] Federal regulatory bodies could endorse SACHRP's recommendation and explicitly permit researchers and IRBs to rely on SACHRP's priority list of potential LARs. Alternatively, state legislatures that have not already done so could draft their own priority lists for LARs for research, eliminating the need to rely on lists derived from laws pertaining to medical treatment. Medical treatment laws define consent capacity with reference to medical decisions, not research decisions. They do not address the task-specific nature of capacity to consent to various research protocols. In addition, legislation drafted for a clinical context assumes that the choices offered to LARs are in the patient's best interest, an assumption that does not necessarily apply to participants in research settings.[299]

* * *

Research including participants with impaired consent capacity presents challenges to researchers, IRBs, institutions, and regulators. Rapidly advancing neuroscience provides an opportunity to revisit these perennial challenges.

Many federal advisory bodies have offered recommendations and proposed guidance to allow research to move forward while protecting potentially vulnerable participants, but no specific regulations have been promulgated. The Bioethics Commission acknowledges the efforts of these previous bodies and the challenges to implementing new policies. In addition, the Bioethics Commission notes the progress made through research to better understand aspects of consent capacity and urges researchers to continue this work to further advance our understanding of the conceptual, legal, ethical, and neurological components of consent capacity.

Better evidence and a clearer analytic synthesis—which integrates the many components of understanding a person's ability to consent in various contexts—are needed to facilitate progress of ethical research that protects participants and seeks to ameliorate the disorders that contribute to impairments in consent capacity. The recommendations outlined in this section illustrate four specific areas of improvement that will move neuroscience forward in an ethically responsible manner and which have the potential—if implemented—to pave the way for moving the national conversation beyond impasse. The Bioethics Commission therefore encourages action in light of, rather than in spite of, justifiably grave concerns and remarkable complexity.

SECTION 4. NEUROSCIENCE AND THE LEGAL SYSTEM

The brains of criminals have captured the public's imagination for centuries. In 1871, when convicted murderer Edward Rulloff was executed for his crimes, scientists acquired his brain to attempt to discover the neurobiological underpinnings of his wrongdoing.[300] More than 200 years later, in 2013, the Public Broadcasting Service aired the series *Brains on Trial*, in which actor Alan Alda took the audience through a series of neurotechnologies and techniques, demonstrating how they might be used in an actual criminal trial.[301] Advances in neuroscience offer a better understanding of human behavior, and the potential for improved policymaking, increased accuracy, and decreased errors in advancing justice. The application of neuroscience to the law also raises concerns—some real and some imagined—about scientific reliability, misapplication and overreliance on a developing science, conceptions of free will, mental privacy, and personal liberty.

Neuroscience has a variety of potential applications to the legal system and already is employed in many relevant contexts, including increasingly in criminal law (Figure 2). Prosecutors and defense attorneys use neuroscience evidence in criminal proceedings to support propositions concerning, for example, competency to stand trial, mitigation of criminal responsibility, and predicting future dangerousness. Parties also use neuroscience evidence in the civil context to provide objective evidence of "invisible" injuries, such as toxic exposure, pain, and suffering. Policymakers have invoked neuroscience to advocate for legislation and reform; scholars have advocated using neuroscience to address biases in legal decision making; and researchers and even some commercial entities have introduced novel uses of neuroscience for investigative purposes.

Neuroscience research holds promise for improving our understanding of human behavior, motivation, intention, and action. However, such research remains in its infancy, and to what extent neuroscience will shape our understanding of these crucial aspects of human behavior is unclear. Moreover, substantial practical limitations constrain what neuroscience is likely to tell us about why *particular* individuals behave in a specific way; thus, neuroscience might offer greater utility for guiding policy decisions, rather than helping to resolve individual criminal or civil cases.

Nonetheless, neuroscience remains poorly understood by the public, attorneys, and judges.[302] Ensuring the ethical application of neuroscience in the legal and policymaking arena requires substantial public and legal education. The Bioethics Commission seeks to discern what neuroscience can and cannot contribute to our legal system now and in the near future and to facilitate its ethical and scientifically credible use.

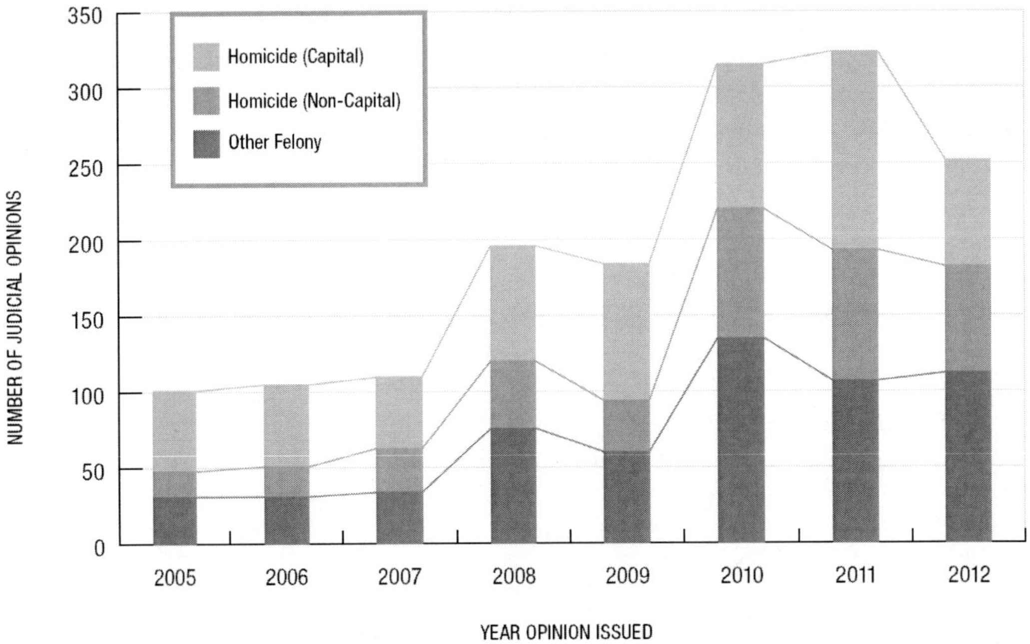

A total of 1800 judicial opinions (majority, plurality, concurrence, dissent) issued during 2005–2012 were included. Graph and analysis based on 1586 majority and plurality opinions only. Source: Farahany, N., Database 2014. On file at Duke University.

Figure 2. Number of U.S. Judicial Opinions Discussing Neurological or Behavioral Genetics Evidence Used by Criminal Defendants, 2005-2012.

Although neuroscience might help us achieve more accuracy in decision making and better policies for trials and sentencing, it does not change the normative or moral questions that the law seeks to answer. Law is a social institution, built on norms developed and instituted by society. Even though neuroscience might guide normative assessments, it cannot solely define them.

Ethical Analysis

Use of neuroscience in the legal realm—from the criminal courtroom to legislation and policymaking—warrants ethical analysis. The potential value of neuroscience to improve decision making accuracy and advance justice must be reconciled with the potential for exaggeration, hype, and premature application of scientific evidence and concepts that are not yet validated, well-understood, or interpreted accurately. Ensuring scientific reliability, scientific literacy among decision makers, and engagement by credible neuroscientists and the public will contribute to increasing neuroscience's value and decreasing unwarranted hype.

Advancing Justice

Neuroscience has the potential to advance justice by increasing accuracy in legal decision making and policy development. A deeper understanding of the human brain, cognition, and behavior on both individual and societal levels might help tailor policies and sentences, determine guilt and innocence, evaluate blameworthiness, and predict future behavior. For example, evidence of brain abnormalities might help determine whether a criminal defendant is competent to stand trial.[303] Neuroscience evidence might contribute to a jury's determination of guilt or innocence, by helping jurors understand a defendant's mental state, intent, or voluntariness of action.[304] A deeper understanding of the development and capacity of the adolescent brain might help formulate policies about the sentences that young adults and adolescents should receive.[305] Neuroscientific techniques like brain imaging might help detect juror bias or determine the reliability of eyewitness testimony.[306] Overall, neuroscience might contribute to more accurate decision making and fairer outcomes. Justice requires that we use empirical evidence, including neuroscience, to strengthen the decisions made in these central civic and political realms.

Enhancing justice by using neuroscience evidence is especially important because of the potentially severe and far-reaching consequences of legal and policy decisions. In the criminal context, punishment can involve deprivation of liberty by imprisonment or the death penalty in some jurisdictions. Such severe consequences warrant particular attention to improving the accuracy of conviction and sentencing. In addition, because legal practice is based on a system of precedent, the application of neuroscience in one courtroom can affect its use in other courtrooms for years to come. In the civil context, courts are clogged with lawsuits that involve disputes over subjective factors like pain and intention, and millions of dollars are spent each year litigating and settling these cases. Neuroscience research efforts to understand and measure these subjective factors can help clarify them and should continue. Interdisciplinary collaboration between neuroscientists and legal decision makers could help advance the cause of justice.

Mitigating Hype

The ability of neuroscience evidence to solve legal and normative questions neatly and cleanly is often exaggerated and hyped. This hype can lead to unwarranted and excessive influence on legal decision makers like judges and jurors.[307] Scientific hype in the media or scientific claims that have not been borne out through replication and verified by the scientific community at large can distort public perception. When legislators rely on hyped scientific claims and unverified science to support political agendas, resulting policies and laws can be unjust.

Thus, neuroscientists, legal decision makers, and scholars must address the tension between advocating the use of neuroscience to improve accuracy and advance justice and prematurely urging its use, potentially hindering justice. The responsibility to avoid hype is shared by many stakeholders, including neuroscientists, members of the media, politicians, judges, and the general public. Public education to improve understanding of neuroscience specifically and scientific evidence more generally is essential to enhance stakeholders' understanding of neuroscientific concepts and the limitations of neuroscience within the legal system, and to reduce the potentially negative effects of hype.

Privacy and Cognitive Liberty

Accurate, reliable, and relevant neuroscience evidence can and should be introduced into the courtroom and policymaking to advance accuracy and justice. However, in the future, neuroscience evidence might raise concerns about cognitive liberty and invasion of privacy.

Some scholars claim that neuroscience brings us one step closer to being able to interrogate the brain or "read minds," which could have implications for individual privacy.[308] However, current technology is extremely limited and is incapable of revealing inner desires, psychological states, or motivations.[309] Still, probing the brain through techniques like neuroimaging raises questions about whether inner mental processes deserve more privacy protection than externally observable clues about the mind.[310] Even failed efforts to penetrate the mind can offend a sense of privacy. Unlike advancing justice and avoiding hype—two more immediate ethical considerations associated with neuroscience and the law—protecting mental privacy is a forward-looking concern that neuroscientists and legal decision makers might need to evaluate as technology continues to advance.

Current Use of Neuroscience within the Legal System

Before further speculating about potential future uses of neuroscience within the legal system and predicting related ethical considerations, the current landscape and the ways neuroscience already guides legal decision making should be assessed. Substantial improvements in neurotechnology and scientists' understanding of the brain, behavior, and cognition in the past decade have led to an increase in the use of neuroscientific evidence within the courtroom and other legal proceedings.[311] Neuroscience use within the legal system follows the recent introduction of behavioral genetics into the courtroom, which presented similar promise to add value and accuracy, but also raised similar ethical and practical concerns.[312]

Scholars have argued that neuroscience hype and the fascination with colorful brain images exerts undue influence in legal decision making.[313] The question of whether brain-based evidence has an impact on judges' and juries' decisions is an empirical one. Judges and juries have acknowledged the persuasive allure of brain scans, after mitigating defendants' sentences in light of neuroscientific evidence.[314] In a nationwide survey, almost 200 state trial judges were presented with a hypothetical case in which psychopathy was diagnosed in a convict. About half of the surveyed judges received expert testimony presenting a biological explanation of psychopathy, and the other half did not. Survey results indicated a correlation between inclusion of the biological explanation of neurological disorder with significantly reduced sentence length and an increased number of mitigating factors listed.[315] Additionally, studies have shown that members of the public are more likely to trust diagnosis of a psychiatric condition when brain imaging evidence is presented, as opposed to evidence from psychological assessments.[316] These data indicate that neuroscience evidence can have substantial effects on legal decision making.

Criminal Law

Neuroscience has become an integral part of the criminal justice system in the United States. During the past decade, hundreds of criminal cases have been influenced by neurobiological claims. In 2012 alone, over 250 judicial opinions—more than double the number in 2007—cite the use of neuroscience by criminal defendants arguing their brain made them do it. Already, over 5 percent of murder trials and 25 percent of death penalty trials feature criminal defendants using neuroscience to argue for lesser responsibility or punishment.[317]

With a few notable exceptions, scientists are on the sidelines of these legal developments. Researchers often decry use of poorly substantiated cognitive neuroscience and call for caution regarding its use in law.

Many methodological problems stymie the reproducibility and validity of neurological studies of complex behavioral traits.[318] For example, what does "impulsivity" mean? How is it measured? Is aggression on the sports field the same as criminal aggression? How do we measure the difference? How can we disentangle environmental and neurological effects in our advancing understanding of human behavior? Do studies on behavioral variations within a population tell us anything about the causes of specific individuals' behavior or actions?

Despite these scientific hurdles, the use of neurological evidence in the criminal courtroom is on the rise. Even the gravest of legal decisions—including the question of capital punishment—already have hinged upon neuroscience.[319] Over 1500 judicial opinions issued during 2005–2012 discuss the use of neuroscience by criminal defendants (Figure 3). Almost 40 percent of those opinions pertain to criminal defendants charged with capital murder, and 61 percent involve defendants charged with other serious offenses, including noncapital homicide, assault, robbery, burglary, drug possession, rape, fraud, and theft.[320] Many of these cases include expert evidence about past head or brain trauma, neuropsychological testing, and neuroimaging studies conducted.

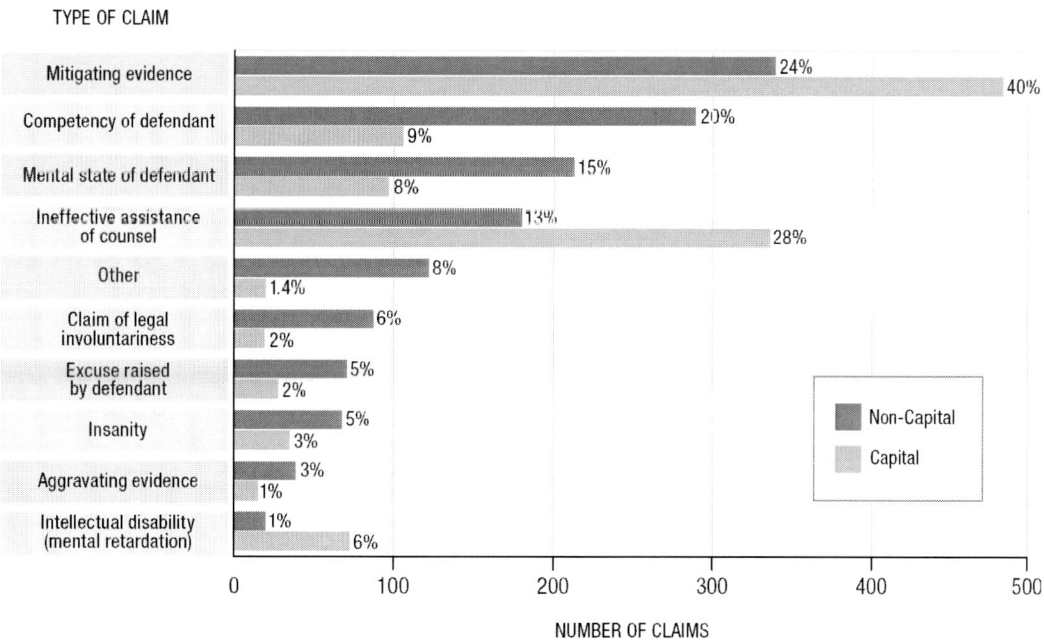

A total of 1800 judicial opinions (majority, plurality, concurrence, dissent) issued during 2005–2012 were included. Graph and analysis based on 1586 majority and plurality opinions only. Source: Farahany, N., Database 2014. On file at Duke University.

Figure 3. Claims Made Using Neurological or Behavioral Genetics Evidence in U.S. Capital and Non-Capital Criminal Cases.

Criminal law involves multiple stages of legal decision making, including competency, trial, and sentencing. Neuroscience evidence has been introduced and employed at various stages. Defendants have argued that they should be held less accountable for their actions or punished less severely because of alleged neurological impairments. Prosecutors have seized upon the double-edged potential of such evidence to indict defendants' characters or underscore future dangerousness. These claims are fueling a reexamination of the criminal justice system.

Competency

The *U.S. Constitution* forbids the trial of a defendant who lacks competency.[321] An attorney can challenge a defendant's competency at any stage of the legal proceedings, from competency to stand trial to competency to be sentenced for a crime. Generally, a defendant is incompetent to stand trial if he suffers from a mental disease or defect that renders him incapable of understanding the charges against him and their potential consequences, or is unable to assist his attorney in his defense.[322] Traditionally, psychological evaluations are used to determine competency. Psychologists conduct interviews or behavioral observations to determine whether accused individuals have the capacity to understand the trial or assist in their own defense.

Attorneys are beginning to rely on neuroimaging techniques to supplement psychological evaluation of competency.[323] Although perhaps only weakly informative, neuroscience can better ascertain the subjective capacities of criminal defendants than existing tools used in law. Judges typically engage in conversational dialogue with defendants and rely upon their perception of defendants' responses to assess competency. Neuropsychological testing, a history of neurological trauma, and neuroimaging might improve such judgments. Imaging techniques can determine whether structural or functional abnormalities exist in individuals' brains that might contribute to a lack of ability to assist at trial.[324] Observation of a physical abnormality alone is insufficient to prove incompetence—the individual's behavior and abilities are relevant. Neuroscience techniques can support competency determinations, especially when fact finders suspect that individuals are lying about their abilities.

Consider for example, David Rothman, a 68-year-old physician, who was charged with conspiring to commit health care fraud against the U.S. government. Before trial and then before sentencing, Dr. Rothman's attorney challenged his competency, calling five different experts to testify about the extensive neuropsychological testing and neuroimaging procedures (including magnetic resonance imaging [MRI], positron emission tomography [PET], and electroencephalogram [EEG]) they had performed on him. The experts believed that Dr. Rothman suffered from a severe loss of insight and comprehension relating to a degenerative brain disorder. The state's counter-expert believed otherwise, based on his limited interactions with Dr. Rothman. After hearing the evidence, the judge agreed with the defense experts, finding Dr. Rothman incompetent to proceed to sentencing. His sentencing was indefinitely suspended.[325]

Trial

Our society, including our legal system, is predicated on understanding and predicting what individuals are thinking, planning, and doing.[326] Although at first glance, the legal system appears primarily concerned with one's actions, an individual's mental state and intention also play a principal role in assigning legal blameworthiness. For example, to be convicted of homicide, one must not only have committed the act of killing someone, but the accused must also have had the requisite *mens rea* (guilty mind)—that is, the intention to have killed the victim.[327] To make those assessments, criminal law has traditionally relied upon observational methods to assess individuals' intentions, such as their testimony, the testimony of others who know them, and the observable circumstances surrounding events. In the future, neuroscience might allow us to achieve more accurate and empirical assessments of individuals' intentions, motives, knowledge, and mental states.[328] Already, defense attorneys have attempted to use neuroscience to try to prove something about individuals' mental states, for example, that they lacked the ability to act with purpose.

Some scholars have argued that neuroscience challenges the "folk psychological" beliefs underlying criminal law: that actions are voluntary and the product of conscious choice.[329] The alternative they propose—that actions arise from unconscious predispositions and decision making over which we have little control—has been largely rejected in criminal law. This is because legal concepts like voluntariness and intentionality are normative and are understood differently by scientists and jurists. This alternative view also does not align well with our subjective experiences of self-directed decision making. Moreover, research has demonstrated that if we approach individuals as responsible actors, they are more likely to act responsibly.[330]

To obtain a criminal conviction, prosecutors must prove beyond a reasonable doubt that a defendant acted voluntarily, with the mental state specified by law. Criminal law grants prosecutors a strong presumption that the defendant acted voluntarily. The defense of involuntariness arises only when the defendant's actions were a reflex or convulsion, or a bodily movement arising from unconsciousness, sleep, hypnosis, or some other factor not in the individual's conscious control.[331] Circumstances like these are believed infrequent in law. However, scientists and philosophers are currently debating whether neuroscience can show that decision making is primarily unconscious, which, if so, would indicate a fundamental mismatch between legal and scientific understanding of the nature of conscious action.[332]

For example, in a 2009 criminal case, a group of four friends gathered for drinks after work. Later that evening, one of the friends drove the intoxicated defendant's truck to take them all home. Upon arriving at his house, the defendant found himself locked out of his home, and returned to the truck where he inexplicably assaulted his friend. The three friends quickly regrouped and started to walk away, but not soon enough. The defendant hopped into the truck and drove it into them, injuring one and killing another. At his trial, a neurologist testified that the defendant's actions were consistent with someone in an automatic, unconscious brain state, taking no purposeful action. The court rejected this claim, finding that the defendant's actions did not appear unconscious or automatic, but instead met the *legal* definition of voluntary, intentional, and purposeful. He was sentenced to 15 years to life in prison.[333]

The mismatch between legal definitions of concepts of guilt and neuroscientific understandings of human behavior has stalled attempts to use neuroscience to challenge voluntariness and even mental states in law. Yet, in quite a few criminal cases, defense attorneys have used neuroscience to argue that criminal defendants lacked the mental state to have committed the crimes. Mental state, like voluntariness, has a precise meaning in law: the defendant's purpose in acting, awareness of the surrounding circumstances, and intent to achieve the resulting consequences.

One problem for defendants in trying to establish their mental state during the commission of the crime is the elapsed time between the crime and later neurological testing. Neuropsychological testing that happens months or years after a crime might have little bearing on the defendant's brain at the time of the crime. Moreover, although neuroscience can inform defendants' general behavioral predispositions—such as an inclination toward impulsivity or aggression—it cannot yet tell us defendants' specific mental states when they engaged in the criminal act.

NEUROSCIENCE EVIDENCE CAN INFLUENCE JURY DECISIONS

Peter Jordan Chiesa owned a parcel of land off the highway. Two neighbors with adjoining land accessed their properties by a dirt road that crossed Mr. Chiesa's land. Over the years, the neighbors clashed regularly about the use, maintenance, and width of the easement. After numerous legal clashes, Mr. Chiesa had the trees pruned on either side of the road. When the neighbors trimmed the trees more, Mr. Chiesa called the police to report that his neighbors were trespassing and indicated that he planned to shoot his neighbors to remove them from his property. He killed two of them.

At trial, the defense argued that Mr. Chiesa—who had no prior criminal record—suffered from brain damage that left him unable to control his emotions and sent him into uncontrollable rages. Computer-assisted tomography (CAT), positron emission tomography (PET), and single-photon emission computerized tomography (SPECT) scans revealed damage in his prefrontal cortex, temporal lobes, and cerebellum—damage that experts claimed would affect his impulse control and temper. His doctors opined that, although Mr. Chiesa was aware of what he was doing when he shot the neighbors, his conduct was driven primarily by impulse, not choice. Despite evident planning—notifying the police of his plan, driving his truck without incident, and aiming the gun at two separate individuals—the jury convicted Mr. Chiesa of the lesser offense of second-degree murder instead of first-degree, premeditated murder.

Source: *People v. Chiesa*. No. CO-47001, 2005 Cal. App. Unpub. LEXIS 10774 (Cal. Ct. App. Nov. 22, 2005).

Sentencing

Neuroscience can help society reexamine why and how we punish individuals for committing crimes. Do we do so because the defendant deserves punishment in proportion to the harm caused to society? If so, does a neurobiological understanding of human behavior undermine retributivism as the basis for punishment? Or do we punish individuals who commit crimes to protect society against dangerous criminals? If so, would this goal be better served by focusing on rehabilitation and reintegration into society for those who commit crimes?

In the majority of criminal cases where neuroscientific evidence has been introduced, it has been used to challenge how severely a defendant should be punished.[334] This evidence often arises in the context of the separate sentencing hearing that defendants receive when convicted of a murder where the death penalty is being sought, to aid the jury in deciding between recommending the death penalty or life in prison without the possibility of parole.

The results for criminal defendants have been mixed. Even in cases with truly chilling facts, courts have held that an attorney for a capital defendant must investigate a reasonable likelihood of a brain abnormality, or risk being found constitutionally ineffective as counsel.[335]

For juvenile offenders, cognitive neuroscience has strongly influenced the recent constitutional prohibitions on execution or sentencing to life imprisonment without the possibility of parole.[336] In 2005, the U.S. Supreme Court ruled that persons younger than 18 years of age at the time of the crime could not receive the death penalty, taking note of an amicus brief that argued that juveniles lack fully matured brains, character, and sense of responsibility.[337] The Supreme Court revisited the question in 2010 and 2012.[338] Citing favorably from briefs submitted by the American Psychological Association and the American Medical Association, the Court held that life imprisonment without the possibility of parole for nonhomicide crimes committed by adolescents is unconstitutional, and that for homicide crimes, the sentencing must be individualized and consider the maturity of the offender.[339]

Prediction

Using neuroscience has a potential double edge for criminal defendants. In some instances, neuroscience has enhanced rather than mitigated a defendant's punishment. Courts have at times regarded defendants' neurological predispositions as aggravating sentencing factors or circumstances.[340] If defendants successfully prove that neurological abnormalities contributed to their criminal conduct, courts might regard their brains as *too* broken or *too* dangerous to have at large, even *if* that demonstrates they are somehow less culpable.[341]

Recidivism prediction (i.e., predicting whether an individual will commit another crime) is used throughout the criminal justice process, including in determinations of bail, sentencing, probation, parole, and treatment program assignments. Methods to predict recidivism include clinical observation and measurement of risk factors, such as age at incarceration, age at release, criminal history, drug use, and social support.

Impulsivity—the persistent lack of restraint and consideration of consequences—is one of the most widely studied risk factors for recidivism. Neuroscience researchers studied the anterior cingulate cortex (ACC), a brain region associated with impulse control, in an experiment attempting to predict future rearrest of previous criminal offenders as a measure of recidivism.[342] Damage to the ACC in humans can lead to changes in inhibition, apathy, and aggression. Although other brain regions also play a role in impulse control, research reveals that the ACC is the most robustly engaged region during impulsive activity.[343]

Another study used functional magnetic resonance imaging (fMRI) scans of hundreds of prisoners to ascertain features that distinguish psychopaths from others.[344] Studies report that individuals diagnosed with psychopathy compose a large percentage of the prison population and are more likely to reoffend.[345] Establishment of biological markers that indicate psychopathy and greater likelihood of recidivism could lead to tailored prison sentences. Some argue that it is unethical to assign longer sentences in response to the same crime for those who are biologically predisposed to reoffending.[346] However, others contend that neuroscience might lend more accuracy to existing methods for predicting recidivism, an established practice in the criminal justice system for making decisions about bail, parole, and sentencing.[347]

> "We use predictions all the time now in criminal law and elsewhere. If we're going to use predictions, and we already think that they are proper to use, I can't imagine what the argument looks like for doing it less well, as opposed to better."
>
> Morse, S., Ferdinand Wakeman Hubbell Professor of Law, Professor of Psychology and Law in Psychiatry, University of Pennsylvania Law School & School of Medicine. (2011). Current Issues in Neuroscience and Neuroimaging. Presentation to the Bioethics Commission, February 28. Retrieved February 2, 2015 from http:// bioethics.gov/node/197.

Alternatives to Sentencing

As neuroscience improves our understanding of the brain and neurological correlates of criminal behavior, alternatives to traditional sentencing and punishment are being proposed and implemented. Several kinds of so-called "treatments" for criminal behavior are already in use, including chemical castration for sex offenders and drugs to combat addiction. Other kinds of regimens are envisioned, such as drugs, psychosurgery, and deep brain stimulation (DBS), to treat impulsivity or aggression.[348]

"Treating" criminal behavior as an alternative to imprisonment or a condition of release raises concerns about the safety and efficacy of these treatments, and about the coercive nature of such options. For example, Depo-Provera®, the female contraceptive injection that is used to chemically castrate male sex offenders, has not been tested in men—but this off-label use by clinicians is not prohibited by the U.S. Food and Drug Administration (FDA).[349] In addition, in some states, chemical castration of sex offenders is mandatory, whereas in others, it is presented as a voluntary alternative to prison or a condition of release. But questions surround whether such a choice can ever be truly voluntary. As neuroscience advances and more treatment options are envisioned—options that might seem like appealing alternatives to a potentially ineffective incarceration system—these outstanding ethical questions must be addressed.[350]

Civil Law

Neuroscience evidence also has been used in the civil law context. Civil law involves suits between private parties that usually involve monetary compensation (as opposed to criminal law cases, which are brought by the government and typically involve such punishment as imprisonment). Workers' compensation, personal injury, and disability determinations are examples of civil litigation, and they often hinge on establishing "invisible injuries," such as exposure to toxins or the presence or extent of pain and suffering, making them ripe for the introduction of neuroscience evidence.[351] Often, pain can be difficult to establish convincingly in civil cases, and neurological testing and evidence potentially could help evaluate the veracity of claims of pain, including emotional pain or suffering.[352]

In contract law, neuroscientific evidence has been proposed for intent verification, that is, to measure an individual's brain activity at the time of contract initiation to determine what the individual intended to agree to and expected to receive in return.[353] For example, in one case, a contract for the sale of land was voided because evidence—including MRI data—demonstrated that one of the contracting parties lacked the mental capacity to enter into contracts.[354] Even in child custody cases, neuroscience has been introduced to establish parents' fitness or neurological evidence about a child's development to determine the best interests of the child.[355]

Accuracy and Errors

Studies are under way to determine whether neuroscience can be used to improve the accuracy of legal decision making and to decrease errors that arise from bias or limitations of human perception. Research has revealed cognitive biases among jurors and judges. For example, implicit bias by jurors based on race and ethnicity is a common concern in U.S. courts.[356] Implicit bias encompasses attitudes and stereotypes that affect our judgments and decisions but do so unconsciously without an individual's awareness, intentionality, or control. Because these biases reside deep within our subconscious, they are usually not accessible to individuals through self-reflection. Studies have revealed implicit biases in the courtroom from previously unexamined factors, for example, differences in how severely a judge sentences a defendant on the basis of whether the sentencing occurs before or after the judge has had lunch.[357]

Current strategies to mitigate implicit bias in the courtroom include acknowledging the existence of implicit bias, routinely examining thought processes and decisions to check for possible bias, and increasing exposure to stereotyped group members.[358] Jurors are unlikely to recognize or reveal their own implicit biases. Neuroscience could help by identifying areas of the brain associated with race-related bias and perhaps eventually using neuroimaging techniques to detect juror bias and correct for it. However, concern exists about the feasibility and ethical constraints of using neuroimaging on all jurors. Techniques like this might be most useful in cases where juror bias is being questioned.

Neuroscience also has been used to challenge and reveal limitations in eyewitness testimony whether through suggestibility that arises from lineup procedures or the limitations of human perception. Ongoing studies are attempting to link brain activation patterns to facial recognition.[359] Using neuroimaging techniques to help witnesses accurately identify offenders is a long way off, but if it becomes possible in the future, it could help to improve accuracy and make just outcomes more likely.

Policymaking

Legislators sometimes rely on neuroscience data and concepts to advocate for their agendas. For science, including neuroscience, to guide just and fair policies and laws, data must be accurate and reliable. Neuroscience might hold greater near-term promise in policymaking than it does in criminal and civil cases. The practical limitations that constrain effective use of neuroscience in the courtroom—including the fact that brain differences *across* a population do not explain the behavior of any *particular* individual in the population—might matter less in policymaking than adjudicating trials. To the extent that neuroscience can help us at a group level to understand how and why people behave as they do, it might provide relevant empirical evidence for new and better social policies. However, scientific hype in the media or scientific claims that have not been borne out through replication and verified by the scientific community at large serve only to distort public perception. If legislators rely on hyped scientific claims and unverified science to support political agendas, resulting policies and laws will be unjust.

For example, as described above, the U.S. Supreme Court has relied on neuroscience evidence in decisions that led to substantial changes in juvenile justice around culpability and punishment of adolescent criminals. In developing policy, lawmakers might draw from evidence concerning impaired judgment of adolescents compared with adults, increased emotional reactivity among adolescents, and their increased likelihood to engage in risky behaviors in the presence of peers despite knowing better.[360]

Scientific evidence should be considered collectively when forming policy decisions. Policy developments based on selected research results or scientific evidence that has not yet been replicated and verified are unlikely to be fair and just. For example, in the 1990s, hyped neuroscience research results led to widespread publicity of a phenomenon called the Mozart effect—an increase in intelligence associated with listening to classical music. The original study demonstrated that college students' performance on spatial reasoning tests improved for 10 to 15 minutes after listening to a Mozart sonata, which translated into temporarily increased IQ scores during that same period.[361] Sustained media interest in the Mozart effect contributed to its false association with children and infants well after the publication of reports failing to replicate the original results.[362] Consequently, in 1998, the Georgia state legislature allocated taxpayer money to a policy ensuring that all newborn infants received a recording of classical music.[363] In this case, unverified scientific results combined with unchecked media hype led to misinformed policymaking and a potential misallocation of taxpayer money.

Disagreement about neuroscience research, findings, and interpretation can have a major impact on the law. Policymakers will likely use neuroscience more frequently as it advances. When mature and verified science is used and interpreted correctly, it can have a positive impact on the law, grounding legislation in evidence. However, neuroscience evidence supporting claims that are not yet fully mature or verified can affect the law prematurely and unjustly.

Interrogating the Brain

Courts and police do their best to ascertain the thoughts, intentions, and desires of criminal defendants. Some argue that neuroscience might eventually allow us to be able to interrogate the brain, threatening a deeply held sense of privacy.[364] But today, and in the foreseeable future, neuroscience does not enable us to read minds. Technology remains extremely limited and cannot reveal the true inner desires, psychological states, or motivations that are worthy of the term "mind-reading."[365]

However, novel neuroscience techniques might soon reveal (with a cooperative witness) whether an individual recognizes a face or an object, possesses knowledge relevant to a legal proceeding, is lying or telling the truth, or even allow reconstruction of the visual imagery seen at the time of the crime.[366] These techniques raise crucial questions, including whether "inner mental or neural processes" deserve more privacy protection than external or behavioral elements such as words and actions, and if so, whether convicting criminals is a sufficient justification for violating that privacy.[367]

If we could accurately interrogate the brain, with a high degree of reliability, then just as DNA evidence has helped to exonerate many wrongfully accused and convicted individuals, so too might neuroscience offer greater accuracy and insights to improve our laws and policies. We should be open to the possibilities that neuroscience can bring, while ensuring the progress and responsible application of neuroscience to the legal system and policymaking.

Scientists disagree about the accuracy of functional brain imaging technologies at issue here, including what inferences can accurately be drawn from observed brain activity. While researchers debate the meaning and utility of these neuroscience techniques, some commercial entities have already started marketing technologies to interrogate the brain.

From simple consumer-based devices that detect electrical activity in the brain and correlate it with attention and relaxation, to complex fMRI scans to detect brain activity associated with complex thoughts, imagery, and tasks, cognitive neuroscience research is advancing our ability to access and decode basic thought processes in the brain.

Researchers have demonstrated through brain scanning that individuals recognize voices or faces with relatively high accuracy in laboratory settings; however, the results are not accurate enough yet to comfortably rely on this technique in a high-stakes situation.[368] Such research might also demonstrate that a person harbors guilty knowledge—for example, knowledge that only individuals familiar with unreleased details of crimes might have—by measuring physiological responses to recognized stimuli using EEG output.[369] With fMRI and sophisticated computer algorithms, researchers have also been able to roughly reconstruct the visual imagery being observed or imagined by an individual, indicating that perhaps one day society might seek to decode the imagery being visualized by a criminal suspect or an eyewitness.[370]

New lie detection tests, initially commercialized by companies like Cephos Corporation (Tyngsboro, Massachusetts), and No Lie MRI (San Diego, California), use fMRI to analyze the truthfulness of individuals in response to questioning.[371] The premise is that lying is more physiologically taxing than telling the truth, and that measuring blood flow patterns and areas of activation of the brain when answering binary questions can accurately identify lying or truth-telling by individuals.[372] Even though litigants have made several attempts to introduce fMRI-based lie-detection in U.S. court cases, no court has admitted such evidence because of concerns about its scientific reliability. However, some countries are already using similar technologies in criminal investigations. Several reported cases in India involved use of similar technology on witnesses and suspects in criminal cases.[373]

These neuroscience applications raise profound ethical and legal questions. For example, do individuals have a right to mental privacy that safeguards them from being compelled to submit to EEG, fMRI, or other brain-based interrogations? Should eyewitnesses have their memories validated by neuroscience techniques? Do the *U.S. Constitution*, conventions on human rights, treaties, or other legal and social norms offer adequate protection to individuals concerning these technologies? Does the Fourth Amendment of the *U.S. Constitution*, which protects individuals from unreasonable searches and seizures by the government, safeguard individuals against such uses?[374] Does the privilege against self-incrimination protected by the Fifth Amendment of the *U.S. Constitution* protect individuals from being compelled to undergo fMRI-based lie detection?[375] What do we mean by freedom of thought, and to what extent is what we mean consistent with neuroscience findings? As neuroscience and technology move forward, scholars will continue to debate these questions, and the public must keep a close eye on the ethical and societal implications of advances and their applications in the courtroom.

The Value of Neuroscience to the Legal System

It is imperative that we avoid overstatements and hype in discussing how neuroscience can affect the legal system. As one commentator described it, neuroscience will not make the legal system "dry up and blow away."[376] Measured introduction of neuroscientific evidence and concepts, after they are validated, well-understood, and interpreted accurately, is potentially highly valuable. Neuroscience cannot answer central normative questions that are important to society—for example, why we punish criminals and what it means to be a morally responsible or free human being. However, it can provide us with evidence and support a better understanding of the human brain that might guide society's consideration of these and other important moral questions. Valid scientific evidence can support evidence-based and more accurate legal and policy decisions. For example, when fingerprinting analysis was first introduced into U.S. courtrooms, it provided evidence relating to whether a defendant committed the act. It does not provide definitive proof; we use other evidence in addition to determine whether the defendant committed the act. Similarly, neuroscience could help address whether an individual has a mental disease or defect—one part of the test of legal insanity in most U.S. jurisdictions. Although neuroscience cannot answer whether an individual is legally insane—a legal construct to determine which individuals we excuse from blameworthiness—it might get us closer to answering whether an individual satisfies the legal test, or even if that legal test can be improved or modified by a better understanding of human behavior.

As neuroscience progresses, it also can contribute to our understanding of how we reason and make moral decisions. Some scholars argue, for example, that over time, neuroscience will reveal that human behavior can be explained as complex causal chains, rather than self-directed choices arising from free will. They argue that this new understanding of human behavior will not necessarily change the way individuals behave or the way we understand human behavior—people will still have the experience of choosing and of making decisions. However, over time, this neuroscientific understanding could challenge the purpose of punishment in the criminal justice system. These scholars argue that, if moral decisions are just inevitable results of causal chains, then individuals are not blameworthy, and punishment for the sake of retribution is intolerable. In this way, they argue, neuroscience will change the way we think about our legal system. Retributive justice will be replaced by punishment only for the sake of deterrence or opportunities for therapeutic interventions.[377]

Claims that the law will be revolutionized by new evidence are not novel.[378] Even though neuroscience, or any science examining the causes of and contributions to human behavior, can help reveal fallacies in our previously held normative judgments, neuroscience does not inevitably lead us to the answers to fundamental moral questions. For example, even if we can agree that moral decisions result from neurological chain reactions, it does not follow that the individuals who make those decisions should not be held ethically or legally responsible for them. The matter remains of what we as individuals and as a society accept as the standards of ethical and legal responsibility. Ultimately, choices about what law and policy should be are social and ethical choices. These choices can be guided by new discoveries from science, but not answered solely by scientific inquiry.

Some scholars have contended that an account of legal responsibility can be defended independently of the existence of free will and that advances in neuroscience will not undermine the basic assumptions of law.[379] Others have used neuroscience to support a legal theory of free will based on freedom of actions.[380] Although science is pertinent to law by providing information about some causes of (and excuses for) behavior, law fundamentally involves normative questions. Law concerns standards for behavior that provide individuals with reasons for action and inaction—either recognition of why an act is wrong or fear of the consequences of violating the law.[381] The determination of legal responsibility concerns not just psychological facts, but social standards or expectations for what it means to have and fulfill legal obligations to other citizens in the community.[382]

> "Neuroscience, like other research on human behavior and decision making, can help us understand the processes by which we make moral decisions."
>
> Greene, J.D., John and Ruth Hazel Associate Professor of Psychology, Director, Moral Cognition Lab, Harvard University. (2014). Potential Implications of Advances in Neuroscience Research for Ethics and Moral Decision Making. Presentation to the Bioethics Commission, June 10. Retrieved February 4, 2015 from http://bioethics.gov/node/3778.

Several examples illustrate this point. As described above, neuroscience evidence is used in determining individuals' competency to stand trial. Previously, psychological evaluations were the only tool at a court's disposal. Advances in neuroscience now answer questions about competency in a more empirical and potentially more accurate way. The legal elements of competency include an understanding of the charges and an ability to assist in one's own defense. Neuroscience techniques that might identify the cognitive capacities necessary to understand trial proceedings and assist in one's own defense could support more accurate competency determinations. For example, neuroscience can address whether an individual has the capacity for long- and short-term memory or whether they have the capacity for sound judgment and reasoning.

Developments in neuroscience likely will continue to refine our understanding of what acting voluntarily or with intentionality means in criminal law. New advances in diagnoses of neurological disorders could provide evidence of involuntary movements, like seizures, which could be used to argue that someone acted involuntarily.[383] Similarly, advances in neuroscience might provide greater clarity about neurological disorders or psychological conditions that cloud consciousness, which might shed light on such symptoms as dissociation and bear on whether behavior was voluntary.[384] Advances in understanding how intoxication and addiction affect psychological states could provide evidence to help jurors differentiate when a defendant has acted recklessly (i.e., with conscious disregard of a substantial and unjustifiable risk of injury to others) rather than negligently, which does not require conscious awareness of a risk.[385] Neuroscientific evidence might supplement the objective circumstantial evidence being used to determine whether a crime was committed in the heat of passion. Advances in the neuroscience of addiction and mental illness also could be relevant to the legal process of diverting those who have committed crimes into treatment programs, perhaps by revealing evidence of who is most likely to be receptive to treatment.[386]

Advances in neuroscience also could help us better understand how incarceration affects cognitive function and whether it is successful at deterring future crime. Neuroscience research can help support better, more practical, and more humane policies. For example, research indicates that solitary confinement, which involves isolation and restriction of environmental and social stimuli, can cause severe psychiatric harm and negatively affect cognitive functioning. Solitary confinement can cause psychological harm among individuals with no history of mental illness and can exacerbate preexisting disorders.[387] Experiments with animals and humans have revealed the effects of isolation and stimuli deprivation on brain degeneration but, because of challenges in working with prison populations, little neurobiological evidence is available of structural brain differences among prisoners confined to solitary confinement. Research on super-maximum (commonly known as "supermax") prisons also demonstrates deleterious effects on prisoners' mental health.[388]

Studies reveal that prisoners are substantially more likely than the general population to have a mental illness or a traumatic brain injury.[389] Self-harm incidents are also more likely among prisoners with a serious mental illness.[390] More research is required on how to treat mentally ill prisoners, how to prevent suicide attempts, and the association between mental illness and recidivism.[391] More empirical neuroscience research will help continue to build understanding of incarceration systems and guide development of better policies.[392] Research about the effects of incarceration on prisoners' brains and on the prevalence of mental illness in prison also might slowly work to affect society's views of criminal justice practices and the ethics of incarceration.[393]

Challenges of Applying Neuroscience to the Legal System

Despite the real and potential value of neuroscience application to the legal system, the transition, translation, and interdisciplinary bridge is not without its challenges, both ethical and practical. Overstatements about the potential value of neuroscience and haste to apply it to the legal system and policy development can lead to misinterpretation and misapplication of scientific information. Drawing legal conclusions on the basis of unreliable science, especially when those conclusions involve the deprivation of an individual's liberty, represents a miscarriage of justice.

Practical Challenges

Although neuroscience can help improve the accuracy of determinations about individuals' mental states, of note, evidence of defendants' *current* mental state does not necessarily pertain to their mental state at the time of an alleged crime. The most relevant judgments about a person's state of mind are unavoidably retrospective.[394] In addition, neurological differences only matter insofar as they correlate to the behaviors of interest. For example, differences in the brains of adolescents and adults do not, on their own, establish that adolescents are less rational. Rather, the association of brain data to behavior, such as irrational behavior, determines the relevance of neuroscientific discoveries to law. In cases such as the adolescent sentencing cases described above, advances in neuroscience supplement or confirm the conclusions that can be drawn from behavioral evidence. Furthermore, the law depends on normative judgments about what it means to be legally responsible. Neuroscience alone cannot determine whether adolescents are reasonable enough to face the death penalty. This determination is not a matter of fact, but a matter of what responses are appropriate for society to employ in light of illegal acts.[395]

Although laboratory studies demonstrate varied potential applications of neuroscience discoveries to the legal setting, many of these discoveries are not ready for use in the courtroom. Studies often have very few participants, many drawn from undergraduate student populations—a sample that does not necessarily represent either the population at large or the defendants to whom the studies are intended to apply.[396] How to apply research results to individuals in a courtroom setting also is unclear, because scientific findings are typically averaged and statistically grouped.[397] Because greater variability can exist between two individuals within a group than between individuals in different groups, determining, for example, which statistical group a defendant falls into can prove challenging.[398] This is commonly referred to in the literature as the general to individual (or group to individual) "G2i" problem.[399]

The G2i problem is a perennial one for courts and represents a fundamental mismatch between population-based science and the individualized determination in law. The goal of science is often to make statistical correlations—for example, neuroscientists typically compile fMRI data from multiple research participants and correlate brain activation patterns with specific cognitive states. But this aggregate information might not accurately reflect an individual's pattern of activity.[400] For example, in one study, aggregate data revealed activation of several brain regions during a deception task, but researchers could not accurately predict deception in all individual research participants.[401] Courts, however, typically are concerned with the individual in front of them rather than aggregate data.[402] When neuroscientists testify as expert witnesses, they must either extrapolate aggregate data to the individual in question, or they can present group data and allow jurors and judges to make inferences and draw legal conclusions about the individual in question. Each of these options is imperfect, and which option an expert employs varies widely across courts.[403]

Ethical Challenges

Several ethical challenges are involved in the application of neuroscience to the legal system. Some ethical concerns are real and applicable today, on the basis of current uses of neuroscience within the legal system. Others, however, are forward-looking concerns involving questions that might arise in the distant future when neuroscience is further advanced.

One contemporary ethical consideration involves treatment of incarcerated individuals and questions of cognitive liberty. In the United States, executing an individual who is determined to be legally insane is unconstitutional; thus, inmates on death row can challenge their competency for execution.[404] At issue, and working its way through courts in several jurisdictions, is the question of whether a death row inmate can constitutionally be forcibly medicated to render him competent to be executed.[405] These questions and the consequences of the court decisions have profound ethical import.

Developments in neuroscience could also provide alternative forms of punishment and rehabilitation for criminals. The prefrontal lobotomy was used for approximately a quarter of a century to mitigate antisocial behavior caused by mental illness.[406] A 2005 study reported that DBS can diminish aggressive behavior; other brain regions associated with criminality might be susceptible to improvement through DBS or even less invasive technologies like transcranial magnetic stimulation.[407] Direct brain interventions in criminal behavior by pharmacological or other neuromodulatory methods might proliferate with advances in neuroscience. As discussed above, this neuroscientific medicalization of crime and punishment is especially concerning, given the vulnerability of incarcerated individuals.

Recommendations

Bridging the gap between neuroscience and the law can be difficult. The two disciplines are vastly different from one another, involving different kinds of expertise, assumptions, terminology, norms, and goals. Much of neuroscience aims to make correlations and draw conclusions in the aggregate about human behavior, whereas the law is more concerned with causality and seeks to draw conclusions about individual behavior and motivation. Admitting expert testimony about neuroscience under the rules of evidence can be challenging. In addition, drawing from and interpreting neuroscience in ways that are relevant and probative of important facts, but that do not overly prejudice judges or jurors, is complex. Members of the public, especially ones who will serve as jurors, would benefit immensely from educational resources that help bring high-level neuroscientific concepts into lay terms. Individuals expected to use and interpret neuroscience, including judges and attorneys, would also benefit from greater availability of basic training that helps ease the interdisciplinary transition of neuroscience into the legal decision-making process. Considering the expanding role of the behavioral sciences and neurosciences within the legal system, judges and lawyers must ensure they understand and correctly interpret emerging scientific evidence. Without sufficient training and guidance, lawyers, judges, and jurors cannot effectively assess the evidence and technologies involved in a growing number of legal cases.

Recommendation 10: Expand and Promote Educational Tools to Aid Understanding and Use of Neuroscience within the Legal System

Government bodies and professional organizations, including legal societies and nonprofit organizations, should develop, expand, and promote training resources, primers, and other educational tools that explain the application of neuroscience to the legal system for distribution to members of the public, jurors, judges, attorneys, and others.

Studies reveal the persuasive power of brain scan images and neuroscience information on individual evaluations of the legitimacy of scientific claims.[408] Public education efforts can inform the public and help individuals better understand and interpret scientific claims. For example, Brainfacts.org provides resources designed to educate stakeholders about basic neuroscience principles and societal implications of neuroscience research and its applications.[409] Public education efforts should be responsive to developments in the rapidly changing field of neuroscience.

Specialized training might not prevent novices from succumbing to the allure of neuroscientific explanations, but extended and specific training might be beneficial.[410] Efforts to train lawyers and judges are already under way. For example, the American Association for the Advancement of Science (AAAS) hosts seminars to educate judges on advances in neuroscience and the issues they might encounter as a result of neuroscience developments.[411] Legal professionals demonstrate substantial interest in training sessions. For example, the MacArthur Foundation's Colloquium for Federal Judges on Law, Neuroscience, and Criminal Justice received 135 applications for just 35 openings in 2013 and 2014.[412] Other resources for lawyers and judges include a neuroscience "boot camp" and several curricular materials published by the MacArthur Foundation Research Network, a Neuro-Literacy 101 course hosted by the Integrative Law Institute, continuing education programs, a forensic psychiatry review course sponsored by the American Academy of Psychiatry and the Law, and a publication funded by the National Center for State Courts, the Open Society Institute, and the State Justice Institute outlining research on implicit racial bias and offering strategies to mitigate it in the courtroom.[413]

*

In addition to the broad educational tools discussed in Recommendation 10, relevant bodies also should fund and conduct specific research and report results regarding use of neuroscience evidence in making important legal and policy decisions. Organizations and government bodies also should publish reports that address the challenges and limitations of neuroscience's application to the legal system.

Recommendation 11: Fund Research on the Intersection of Neuroscience and the Legal System

Relevant bodies, such as the National Academies of Science, the U.S. Department of Justice, the National Institute of Justice, and the Social Security Administration, should support comprehensive studies of the use of neuroscience in legal decision making and policy development.

In response to the surge in DNA evidence used in criminal investigations and trials and calls from the scientific and legal communities, the National Academies National Research Council's Committee on DNA Forensic Science produced two detailed reports to address controversy and to answer questions about the use of DNA forensic evidence in criminal cases.[414] These reports proved invaluable to the public and other stakeholders affected by use of DNA evidence. The National Academies similarly issued powerful reports on use of polygraph examinations, recommending against federal government use of polygraphs for screening prospective or current employees to identify spies or other national security risks because of the unreliability of the tests.[415] Similar reports and training materials on the use of neuroscience in law should be developed. These should address the limitations, challenges, and potential for the use of neuroscience in the courtroom and of neuroimaging techniques for investigative purposes.

In addition, untapped resources for empirical work in this area abound, for example, through such government agencies as the Social Security Administration that collect and process data surrounding medical claims in administrative legal proceedings. These agencies might benefit from supporting new studies to aid understanding of the neurobiological bases of pain and disability to facilitate accuracy in claim processing and arbitration.[416]

*

Neuroscience can add value to legal decision making and policy development—strengthening normative conclusions, lending accuracy and scientific evidence to often intangible information, such as mental state and intent, and potentially weeding out problems of bias and faulty memory. However, neither neuroscience nor any other science will inevitably overturn existing norms undergirding the legal system or fundamentally alter our notions of blameworthiness and responsibility. To maximize the value that neuroscience has to offer, scientists, the media, and legal decision makers must avoid hype. Unrealistically high expectations for new science and technology can lead to a loss of trust when those expectations are unmet.[417]

Recommendation 12: Avoid Hype, Overstatement, and Unfounded Conclusions

Neuroscientists, attorneys, judges, and members of the media should not overstate or rely too heavily on equivocal neuroscientific evidence to draw conclusions about behavior, motivations, intentions, or legal inferences.

As scholars have noted, "[p]eople studying the ethical, legal and social implications of neuroscience have to walk a tightrope."[418] The ethical implications of potential technologies must be considered before those technologies are used widely. But scholars have been criticized for putting the cart before the horse—puzzling through potential implications of a technology that is not ready for valid and reliable use creates the expectation that it works.[419]

In addition, when neuroscience evidence that is unreliable or has not yet been validated and is not ready for application is introduced into the legal system, justice is threatened. Neuroscientists and courts must exercise caution to maximize the value of neuroscience to the legal decision-making process.

*

As attorneys introduce more neuroscience evidence into the courtroom, and advocates use neuroscience to influence policy, neuroscientists should engage with the process, consider potential legal applications of their work, and seek to engage with legal and policy decision makers to ease the translation. Neuroscientists can play a principal role in assisting judges and jurors to determine the appropriate interpretations of neuroscientific evidence.[420] Judges need to understand how neurotechnologies work and what their limitations are, and the limitations of data translation to the courtroom, such as the general to individual problem.[421]

Recommendation 13: Participate in Legal Decision-Making Processes and Policy Development

Neuroscientists should participate in legal decision-making processes and policy development to ensure the accurate interpretation and communication of neuroscience information.

Academic institutions, neuroscience and ethics professional organizations, and science policy organizations can play a role in increasing responsible engagement of neuroscientists with legal decision-making processes and policy development. They can offer educational materials and training resources that describe opportunities for engagement, help neuroscientists understand legal applications for their work, and develop communication skills to bridge language and methodological gaps between the two fields.

Some ways that neuroscientists can engage with these processes include acting as expert witnesses, acting as consultants for a legal team, or helping to file amicus briefs to courts.[422] Training resources for neuroscientists acting as expert witnesses should emphasize that neuroscientists need to understand certain points, including the different ways that science and the courts approximate truth; that opposing attorneys will attempt to expose flaws in their data and in their testimony; that being an expert witness entails answering specific questions as opposed to delivering a lecture; that judges are responsible for determining whether their expert testimony is admissible; and that legal jargon is different from scientific jargon and that bridging that gap can be challenging.[423]

* * *

Neuroscience offers a variety of current and potential future applications in legal decision making and policy development. Although neuroscience will not change the fundamental normative inquiries at the heart of the legal system, it provides tools to help address them with more evidence and increased accuracy, and has the potential to advance the cause of justice and the rights of the accused and the incarcerated. Practical challenges to its importation into the legal system persist, and ethical challenges, both real and imagined, must be considered. Education is essential to combat challenges of hype and reliability and address ethical concerns head-on. Neuroscientists should engage in and lend their expertise to the legal process. The Bioethics Commission urges stakeholders to consider the potential impact of neuroscience on the legal system and policymaking, continue efforts to educate decision makers and others, and strive for reliability and fair and just application.

CONCLUSION

The President charged the Bioethics Commission to examine neuroscience research ethics and the societal implications of applications of neuroscience research as part of the BRAIN Initiative. The Bioethics Commission addressed this charge in two parts. This second part, *Gray Matters*, Vol. 2, takes an in-depth look at three of the most controversial topics at the intersection of neuroscience and society that have captured the public's attention—cognitive enhancement and other neural modification, consent capacity, and neuroscience and the legal system. In addition, this volume proposes one overarching recommendation that pertains to all funders associated with the BRAIN Initiative.

Recommendation

In this report, the Bioethics Commission calls for research on a number of critical topics. Such research requires adequate support, including funding, personnel, and other resources. As a White House Grand Challenge, the BRAIN Initiative is uniquely positioned to establish and support efforts that bring together diverse expertise from neuroscience, ethics, law, policy, and other disciplines to advance research and education at the intersection of neuroscience, ethics, and society.

Recommendation 14: Establish and Fund Multidisciplinary Efforts to Support Neuroscience and Ethics Research and Education

The BRAIN Initiative should establish and fund organized, independent, multidisciplinary efforts to support neuroscience and ethics research and education, including the activities recommended in this report.

Financial, practical, and technical support is necessary to investigate important unresolved research and policy questions, many of which are outlined in this report. Organized, independent, multidisciplinary efforts would provide infrastructure to address ethics integration, education, and research. These efforts need not be single brick-and-mortar centers, housed within a particular federal agency or institution. Other configurations can enable stakeholders from an array of public and private entities to collaborate and facilitate research on the ethical and societal implications of neuroscience as the field advances.

Existing initiatives can guide how these efforts might be structured. For example, the HIV [human immunodeficiency virus] Modelling Consortium is a coordinated effort led by representatives from major governmental and nongovernmental organizations. The Consortium identifies questions for further research, facilitates data and information sharing across a diverse set of projects, provides a forum for review of new research, and funds new research.[424] Located around the country, Consortium members meet virtually through periodic meetings and teleconferences. The Human Genome Project Ethical, Legal and Societal Implications Research Program (HGP ELSI) used set-aside funds to support research on related ethical issues. Over the past two decades, a community of interdisciplinary scholars has offered constructive critique of the structure of HGP ELSI and the implementation of its research and education components.[425] The architects of these new neuroscience funded efforts should look to these examples and others to discern successful approaches.

Potential topics that should be addressed by these efforts include the possibility that neuroscience research might contribute to or mitigate stigma and discrimination, be used as evidence in legal proceedings, improve clinical practice, and influence public perceptions of neuroscience and neurological disorders, among others. In addition, these collaborative efforts should support development and dissemination of training materials, guidance, and other educational tools to facilitate informed public debate. Efforts funded by the BRAIN Initiative that integrate neuroscience and ethics will be well-positioned to answer new and remaining ethical questions, consider societal implications of neuroscience research, educate the public, and implement policy recommendations.

* * *

Advances in contemporary neuroscience research have the potential to deliver both individual and societal benefits, including the potential to treat or cure devastating neurological disorders, and to further our understanding of the human brain and mind. Neuroscience advances have captured the public's attention and stimulated scholarly and public debate, fueled by accurate accounts of the science as well as hyped and misinformed interpretations. Three controversies represent some of the most important and provocative topics at the intersection of neuroscience and society. Neural modification, including cognitive enhancement, raises questions about reconciling risks and benefits, ensuring justice, and understanding what it means to be human. Adequately respecting and protecting individuals with impaired consent capacity has presented challenges for decades. Advances in neuroscience and the promise of neuroscience research compel us to reexamine this area—ensuring that those with impaired consent capacity can participate in and benefit from ethical research. Application of neuroscience to legal decision making and policy development offers the potential for more accurate and just outcomes, but also raises concerns about premature use of scientific information, privacy, and moral responsibility. In this report, the Bioethics

Commission seeks to clarify the scientific landscape, identify common ground, and recommend ethical paths forward to stimulate and continue critical, well-informed conversations at the intersection of neuroscience and ethics as the field continues to advance.

APPENDIX I: HISTORY OF MAJOR U.S. POLICY PROPOSALS AND RECOMMENDATIONS ON CONSENT CAPACITY IN RESEARCH

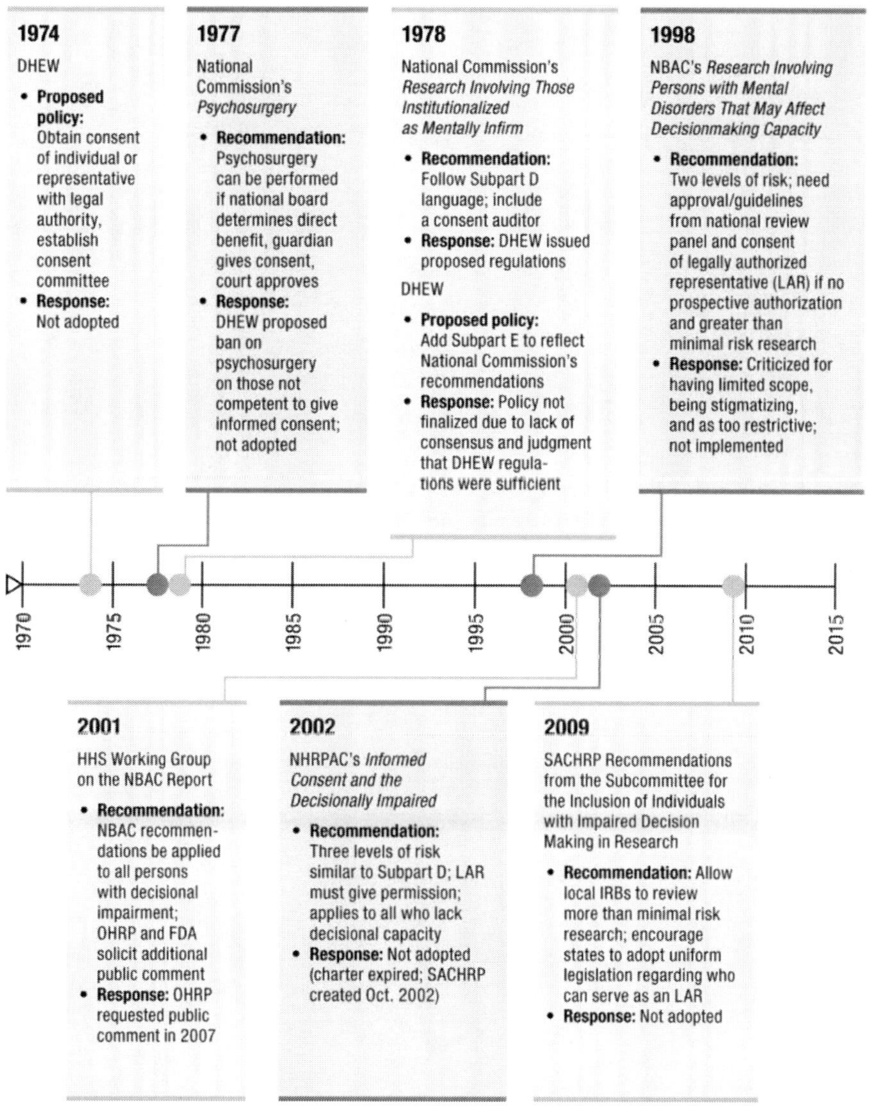

DHEW: Department of Health, Education, and Welfare; HHS: U.S. Department of Health and Human Services (formerly U.S. Department of Health, Education, and Welfare); IRBs: institutional review boards; National Commission: National Commission for the Protection of Human Subjects of Biomedical and Behavior Research; NBAC: National Bioethics Advisory Commission; NHRPAC: National Human Research Protections Advisory Committee; OHRP: Office for Human Research Protections (under HHS); SACHRP: Secretary's Advisory Committee on Human Research Protections; Subpart refers to subparts of *Protection of Human Subjects, HHS*. 45 C.F.R. Part 46

APPENDIX II: GUEST PRESENTERS TO THE BIOETHICS COMMISSION REGARDING ETHICS AND NEUROSCIENCE

Paul S. Appelbaum, M.D.
Elizabeth K. Dollard Professor of Psychiatry, Medicine & Law; Director, Division of Psychiatry, Law, and Ethics; Director, Center for Research on Ethical, Legal & Social Implications of Psychiatric, Neurologic & Behavioral Genetics, Department of Psychiatry, Columbia University

Giorgio A. Ascoli, Ph.D.
University Professor, Molecular Neuroscience Department; Founding Director, Center for Neural Informatics, Structures, and Plasticity, Krasnow Institute for Advanced Study, George Mason University

Nick Bostrom, Ph.D.
Professor, Faculty of Philosophy; Director, Future of Humanity Institute; Director, Programme on the Impacts of Future Technology, University of Oxford

William D. Casebeer, Ph.D. (U.S.A.F., Retired)
Program Manager, Defense Advanced Research Projects Agency

Timothy Caulfield, LL.M., F.R.S.C., F.C.A.H.S.
Canada Research Chair in Health Law and Policy, Professor in the Faculty of Law and the School of Public Health, University of Alberta

David Chalmers, Ph.D.
Professor of Philosophy and Co-director of the Center for Mind, Brain, and Consciousness, New York University; Distinguished Professor of Philosophy and Director of the Centre for Consciousness, Australian National University

Anjan Chatterjee, M.D., F.A.A.N.
Professor of Neurology, Center for Cognitive Neuroscience and Center for Functional Neuroimaging, University of Pennsylvania School of Medicine

Mildred Cho, Ph.D.
Associate Director; Professor of Pediatrics, Stanford Center for Biomedical Ethics, Stanford University

Sohini Chowdhury
Senior Vice President, Research Partnerships, Michael J. Fox Foundation for Parkinson's Research

Miyoung Chun, Ph.D.
Executive Vice President of Science Programs, The Kavli Foundation

Patrick Corrigan, Psy.D.
Distinguished Professor of Psychology, Illinois Institute of Technology

RADM Peter J. Delany, Ph.D., LCSW-C
Director, Center for Behavioral Health Statistics and Quality, Substance Abuse and Mental Health Services Administration

Rebecca Dresser, J.D.
Daniel Noyes Kirby Professor of Law; Professor of Ethics in Medicine, Washington University, St. Louis

Margaret Eaton, Pharm.D., J.D.
Former Research Scholar, Stanford University Center for Biomedical Ethics; Former Lecturer, Stanford University Graduate School of Business and School of Medicine

Martha Farah, Ph.D.
Walter H. Annenberg Professor in Natural Sciences; Professor of Psychology; Director, Center for Cognitive Neuroscience; Director, Center for Neuroscience and Society; Senior Fellow, Center for Bioethics, University of Pennsylvania

Howard Feldman, M.D., F.R.C.P.
Professor of Neurology and Executive Dean, Research; Faculty of Medicine, University of British Columbia

Erik Fisher, Ph.D.
Associate Director for Integration, Center for Nanotechnology in Society; Assistant Professor, School of Politics and Global Studies and the Consortium for Science, Policy and Outcomes, Arizona State University

Paul J. Ford, Ph.D.
Program Director, NeuroEthics Program; Education Director, Department of Bioethics, Cleveland Clinic; Associate Professor, Division of Medicine, Cleveland Clinic Lerner College of Medicine of Case, Western Reserve University

William D. Graf, M.D.
Professor, Departments of Pediatrics and Neurology, Yale University, School of Medicine

Hank Greely, J.D.
Deane F. and Kate Edelman Johnson Professor of Law, Stanford Law School; Professor (by courtesy) of Genetics, Stanford Medical School; Director, Center for Law and the Biosciences; Director, Stanford Interdisciplinary Group on Neuroscience and Society and its Program in Neuroethics, Stanford Law School; Chair, Steering Committee of the Center for Biomedical Ethics

Joshua D. Greene, Ph.D.
John and Ruth Hazel Associate Professor of Psychology; Director, Moral Cognition Lab, Harvard University

Barbara Herr Harthorn, Ph.D.
Director, NSF Center for Nanotechnology in Society; Professor, Department of Anthropology, University of California, Santa Barbara

Freddie Ann Hoffman, M.D.
Chief Executive Officer and Founding Member, HeteroGeneity LLC; Former Deputy Director, Medicine Staff, Office of the Commissioner, U.S. Food and Drug Administration

Steven E. Hyman, M.D.
Founding President, International Neuroethics Society; Director, Stanley Center for Psychiatric Research, Broad Institute of Massachusetts Institute of Technology and Harvard University

Judy Illes, Ph.D.
Professor of Neurology; Canada Research Chair in Neuroethics; Director, National Core for Neuroethics; Faculty, Brain Research Centre, University of British Columbia

Harry Johns
President and Chief Executive Officer, Alzheimer's Association

Deborah G. Johnson, Ph.D., M.Phil., M.A.
Anne Shirley Carter Olsson Professor of Applied Ethics, Science, Technology, and Society Program, Department of Engineering and Society, School of Engineering and Applied Science, University of Virginia

Christof Koch, Ph.D.
Chief Scientific Officer, Allen Institute for Brain Science

Walter J. Koroshetz, M.D.
Deputy Director, National Institute of Neurologic Disease and Stroke, National Institutes of Health

Pat Levitt, Ph.D.
Chair-Elect, Neuroscience Section, American Association for the Advancement of Science; Provost Professor, Department of Pediatrics, W.M. Keck Chair in Neurogenetics, Keck School of Medicine, University of Southern California; Director, Program in Developmental Neurogenetics, Institute for the Developing Mind, Children's Hospital Los Angeles

Herbert S. Lin, Ph.D.
Chief Scientist, Computer Science and Telecommunications Board, National Research Council of the National Academies

Bernard Lo, M.D.
Professor of Medicine; Director, Program in Medical Ethics, University of California, San Francisco

Peggy Mason, Ph.D.
Chair, Ethics Committee, Society for Neuroscience; Professor, Department of Neurobiology, University of Chicago

Helen Mayberg, M.D.
Professor of Psychiatry, Neurology, and Radiology; Dorothy C. Fuqua Chair, Psychiatric Neuroimaging and Therapeutics, Emory University School of Medicine

Robert McGinn, Ph.D.
Professor of Management Science and Engineering; Professor of Science, Technology, and Society, Stanford University

Deven McGraw, J.D., M.P.H., LL.M.
Partner, Manatt, Phelps & Phillips, LLP; Former Director, Health Privacy Project, Center for Democracy and Technology

Alfred R. Mele, Ph.D.
William H. and Lucyle T. Werkmeister Professor of Philosophy, Florida State University

Jerry Menikoff, M.D., J.D.
Director, Office for Human Research Protections, U.S. Department of Health and Human Services

Jonathan Montgomery, LL.M.
Chair, Nuffield Council on Bioethics; Professor of Health Care Law, University College London

Stephen Morse, J.D., Ph.D.
Ferdinand Wakeman Hubbell Professor of Law; Professor of Psychology and Law in Psychiatry, University of Pennsylvania Law School and School of Medicine

Thomas H. Murray, Ph.D.
President Emeritus, The Hastings Center

Ushma Neill, Ph.D.
Director, Office of the President, Memorial Sloan-Kettering Cancer Center; Editor at Large, *Journal of Clinical Investigation*

Helen Nissenbaum, Ph.D.
Professor, Media, Culture, and Communication, and Computer Science; Director, Information Law Institute, New York University

Carlos Peña, Ph.D., M.S.
Director, Division of Neurological and Physical Medicine Devices, Office of Device Evaluation, Center for Devices and Radiological Health, U.S. Food and Drug Administration

John Perry, Ph.D.
Emeritus Professor of Philosophy, Stanford University; Distinguished Professor of Philosophy, University of California, Riverside

Eric Racine, Ph.D.
Director, Neuroethics Research Unit; Associate Research Professor, Institut de Recherches Cliniques de Montréal; Associate Research Professor, Department of Medicine, Université de Montréal; Adjunct Professor, Department of Medicine and Department of Neurology and Neurosurgery, McGill University

Adrian Raine, Ph.D.
Richard Perry University Professor of Criminology, Psychiatry, and Psychology, University of Pennsylvania

Peter Reiner, V.M.D., Ph.D.
Professor, National Core for Neuroethics, Kinsmen Laboratory of Neurological Research and Brain Research Centre, Department of Psychiatry, University of British Columbia

John Reppas, M.D., Ph.D.
Director of Public Policy, Neurotechnology Industry Organization

Nikolas Rose, Ph.D.
Member, Human Brain Project Social and Ethical Division Steering Committee; Professor of Sociology and Head of Department of Social Science, Health and Medicine, King's College London

Bruce R. Rosen, M.D., Ph.D.
Professor of Radiology, Harvard Medical School; Director, Athinoula A. Martinos Center for Biomedical Imaging, Department of Radiology, Massachusetts General Hospital

Adina Roskies, Ph.D.
Associate Professor, Department of Philosophy, Dartmouth College

Pamela Sankar, Ph.D.
Associate Professor, Department of Medical Ethics and Health Policy; Senior Fellow, Leonard Davis Institute of Health Economics, University of Pennsylvania

Marya Schechtman, Ph.D.
Professor of Philosophy, University of Illinois at Chicago

Terrence J. Sejnowski, Ph.D.
Francis Crick Chair, Professor and Laboratory Head, Computational Neurobiology Laboratory, Salk Institute for Biological Studies; Investigator, Howard Hughes Medical Institute; Distinguished Professor, Section of Neurobiology/Neurosciences, University of California, San Diego

Stefano Semplici, Ph.D.
Chairperson, International Bioethics Committee, United Nations Educational, Scientific, and Cultural Organization; Professor of Social Ethics, University of Rome Tor Vergata

Gregory Simon, M.D., M.P.H.
Investigator, Group Health Research Institute; Chair, Scientific Advisory Board, Depression and Bipolar Support Alliance

Steven L. Small, Ph.D., M.D.
Stanley van den Noort Professor and Chair, Department of Neurology; Professor of Neurobiology and Behavior; Professor of Cognitive Sciences; Director, Neuroscience Imaging Center, University of California, Irvine; Professor Emeritus of Neurology, The University of Chicago

Mildred Z. Solomon, Ed.D.
President and CEO, The Hastings Center; Clinical Professor of Anaesthesia, Harvard Medical School

Nicholas Steneck, Ph.D.
Director, Research Ethics and Integrity Program, Michigan Institute for Clinical and Health Research; Professor Emeritus of History, University of Michigan

Serena Viswanathan, J.D.
Assistant Director, Division of Advertising Practices, Bureau of Consumer Protection, U.S. Federal Trade Commission

Anthony Wagner, Ph.D.
Professor of Psychology and Neuroscience, Stanford University

Stephen J.A. Ward, Ph.D.
Professor and Director, George S. Turnbull Center, School of Journalism and Communication, University of Oregon-Portland

John C. Wingfield, Ph.D.
Assistant Director for the Directorate for Biological Sciences, National Science Foundation

Paul Root Wolpe, Ph.D.
Director, Center for Ethics; Asa Griggs Candler Professor of Bioethics, Emory University

David E. Wright, Ph.D.
Director, Office of Research Integrity, U.S. Department of Health and Human Services

David W. Wright, M.D.
Associate Professor of Emergency Medicine; Director of Emergency Neurosciences, Emory University School of Medicine

End Notes

[1] Presidential Commission for the Study of Bioethical Issues (PCSBI). (2014, May). *Gray Matters: Integrative Approaches for Neuroscience, Ethics, and Society*. Washington, DC: PCSBI.

[2] Ferrie, J.E., et al. (2011). Change in sleep duration and cognitive function: Findings from the Whitehall II study. *Sleep*, 34(5), 565-573; Colombo, J., et al. (2013). Long-term effects of LCPUFA supplementation on childhood cognitive outcomes. *American Journal of Clinical Nutrition*, 98(2), 403-412; Taylor, M.P., et al. (2010). Lessons learned on lead poisoning in children: One-hundred years on from Turner's declaration. *Journal of Pediatrics and Child Health*, 47(12), 849-856; Hu, J., et al. (2014). Exercise improves cognitive function in aging patients. *International Journal of Clinical and Experimental Medicine*, 7(10), 3144-3149.

[3] Lucke, J., and B. Partridge. (2013). Towards a smart population: A public health framework for cognitive enhancement. *Neuroethics*, 6(2), 419-427.

[4] Ragan, C.I., Bard, I., and I. Singh. (2013). What should we do about student use of cognitive enhancers? An analysis of current evidence. *Neuropharmacology*, 64, 588-595; Smith, M.E., and M.J. Farah. (2011). Are prescription stimulants "smart pills"? The epidemiology and cognitive neuroscience of prescription stimulant use by normal healthy individuals. *Psychological Bulletin*, 137(5), 717-741; Ilieva, I., Boland, J., and M.J. Farah. (2013). Objective and subjective cognitive enhancing effects of mixed amphetamine salts in healthy people. *Neuropharmacology*, 64, 496-505; Farah, M.J., et al. (2014). Cognitive enhancement. *WIREs Cognitive Science*, 5, 95-103.

[5] Buchanan, A. (2011). *Beyond Humanity: The Ethics of Biomedical Enhancement*. New York, NY: Oxford University Press; Greely, H., et al. (2008). Towards responsible use of cognitive-enhancing drugs by the healthy. *Nature*, 456(7223), 702-705.

[6] Hotze, T.D., et al. (2011). "Doctor, would you prescribe a pill to help me...?" A national survey of physicians on using medicine for human enhancement. *American Journal of Bioethics*, 11(1), 3-13; Larriviere, D., et al. (2009). Responding to requests from adult patients for neuroenhancements: Guidance of the Ethics, Law and Humanities Committee. *Neurology*, 73(17), 1406-1412.

[7] Graf, W., et al. (2013). Pediatric neuroenhancement. *Neurology*, 80(13), 1251-1260.

[8] Labuzetta, J.N., Burnstein, R., and J. Pickard. (2011). Ethical issues in consenting vulnerable patients for neuroscience research. *Journal of Psychopharmacology*, 25(2), 205-210; Grisso, T., and L. Steinberg. (2005). Between a rock and a soft place: Developmental research and the child advocacy process. *Journal of Clinical Child and Adolescent Psychology*, 34(4), 619-627; Appelbaum, P.S. (2010). Consent in impaired populations. *Current Neurology and Neuroscience Reports*, 10(5), 367-373.

[9] *Protection of Human Subjects, Department of Health and Human Services (HHS)*. 45 C.F.R. §§ 46.102, 46.111.

[10] Rusconi, E., and T. Mitchener-Nissen. (2014). The role of expectations, hype and ethics in neuroimaging neuromodulation futures. *Frontiers in Systems Neuroscience*, 8, 1-7.

[11] Jones, O.D., et al. (2013). Neuroscientists in court. *Nature Reviews Neuroscience*, 14(10), 730-736.

[12] Collins, F., and A. Prabhakar. (2013, April 2). BRAIN Initiative Challenges Researchers to Unlock Mysteries of Human Mind. *The White House Blog*. Retrieved February 2, 2015 from http://www.whitehouse.gov/blog/ 2013/ 04/02/brain-initiative-challenges-researchers-unlock-mysteries-human-mind.

[13] PCSBI, *supra* note 1.

[14] Mentions of the brain and related technologies in media headlines have increased exponentially in recent years. See, e.g., North, A. (2014, November 14). Do you want to know about the brain? *New York Times*. Retrieved February 2, 2015 from http://op-talk.blogs.nytimes.com/2014/11/14/do-you-want-to-know-about-your-brain/; Dublkevic, V., Saigle, V., and E. Racine. (2014). The rising tide of tDCS in the media and academic literature. *Neuron*, 82(4), 731-735.

[15] Purcell, R. (2014, July 24). The new normal: How the definition of disease impacts enhancement. *The Neuroethics Blog*. Retrieved February 2, 2015 from http://www.theneuroethicsblog.com/2014/07/the-new-normal-how-definition-of.html?spref=tw. See also, Vincent, N.A., and E.A. Jane. (2014, June 15). Put down the smart drugs – Cognitive enhancement is ethically risky business. *The Conversation*. Retrieved February 2, 2015 from http://theconversation.com/put-down-the-smart-drugs-cognitive-enhancement-is-ethically-risky-business-27463.

[16] Earp, B. (2014, May 13). When is diminishment a form of enhancement? Another twist to the "enhancement" debate in biomedical ethics. *The Neuroethics Blog*. Retrieved February 2, 2015 from http://www.theneuroethicsblog.com/2014/05/when-is-diminishment-form-of.html.

[17] Provigil® is produced by Teva Pharmaceuticals, Frazer, Pennsylvania. Singh, M. (2014, November 17). Are modafinil's brain-boosting benefits hype or science? *NPR*. Retrieved February 2, 2015 from http://www.npr.org/blogs/health/2014/11/17/363840431/are-modafinils-brain-boosting-benefits-hype-or-science; Locke, S. (2014, November 11). How bionic technology will change what it means to be human. *Vox*. Retrieved February 2, 2015 from http://www.vox.com/2014/11/11/7175455/cyborg-ethics-moreno; Clark, V.P. (2014). The ethical, moral, and pragmatic rationale for brain augmentation. *Frontiers in Systems Neuroscience*, 8, 130; Nagel, S.K. (2014). Enhancement for well-being is still ethically challenging. *Frontiers in Systems Neuroscience*, 8, 72.

[18] See, e.g., Underwood, E. (2014, October 22). Neuroscientists speak out against brain game hype. *Science*. Retrieved February 2, 2015 from http://news.sciencemag.org/brain-behavior/2014/10/neuroscientists-speak-out-against-brain-game-hype; Koenig, R. (2014, October 22). Brain-training companies get advice from some academics, criticism from others. *The Chronicle of Higher Education*. Retrieved February 2, 2015 from http://chronicle.com/article/Brain-Training-Companies-Get/149555/; Thompson, C. (2014, October 23). Can video games fend off mental decline? *New York Times*. Retrieved February 2, 2015 from http://www.nytimes.com/2014/10/26/magazine/can-video-games-fend-off-mental-decline.html?emc=eta1&_r=0; Gholipour, B. (2014, October 25). No proof that 'brain training' games work, some experts say. *NBC News*. Retrieved February 2, 2015 from http://www.nbcnews.com/id/56299301/ns/technology_and_science-science/t/no-proof-brain-training-games-work-some-experts-say/#.VHOXE8naLxt.

[19] Cha, A.E. (2014, June 30). Can the secrets to human cognition be found in Lumosity's brain training games? *The Washington Post*. Retrieved February 2, 2015 from http://www.washingtonpost.com/national/health-science/can-the-secrets-to-human-cognition-be-found-in-lumositys-brain-training-games/2014/06/30/b30bf1a2-e73a-11e3-a86b-362fd5443d19_story.html.

[20] Sahakian, B.J., President, and S.E. Hyman., Past-President, International Neuroethics Society. (2014, March 31). Comments submitted to the Presidential Commission for the Study of Bioethical Issues (PCSBI).

[21] Shapiro, R.S., and J. Shapiro-Barr. (2014, July 25). The FDA's informed consent draft guidance: Insights and challenges for IRBs, clinical investigators and sponsors. *The National Law Review*. Retrieved February 2, 2015 from http://www.natlawreview.com/article/fda-s-informed-consent-draft-guidance-insights-and-challenges irbs clinical investig.

[22] Frieden, J. (2014, July 22). FDA issues informed consent guidance. *MedPage Today*. Retrieved February 2, 2015 from http://www.medpagetoday.com/PublicHealthPolicy/ClinicalTrials/46888.

[23] Walker, E.F., and A.T. Ryan. (2014, July 15). Intellectual property from clinical research on neuropsychiatric disorders: What constitutes informed consent? *The Neuroethics Blog*. Retrieved February 2, 2015 from http://www.theneuroethicsblog.com/2014/07/intellectual-property-from-clinical.html.

[24] Capron, A.M., Board Chair, Public Responsibility in Medicine and Research. (2014, March 26). Comments submitted to the Presidential Commission for the Study of Bioethical Issues (PCSBI).

[25] Public Broadcasting Service (PBS). (n.d). Brains on Trial. Retrieved February 2, 2015 from http://brainsontrial.com/.

[26] DiSalvo, D. (2014, October 21). Can neuroscience debunk free will? *Time*. Retrieved February 2, 2015 from http://time.com/3529770/neuroscience-free-will/; Burkeman, O. (2014, November 21). Free will: Are we just slaves to our bodies? *The Guardian*. Retrieved February 2, 2015 from http://www.theguardian.com/lifeandstyle/2014/nov/21/free-will-mind-rule-body-oliver-burkeman.

[27] Greely, H.T. (2014, October 15). Will neuroscience radically transform the legal system? *Slate*. Retrieved February 2, 2015 from http://www.slate.com/articles/technology/future_tense/2012/10/fmri_in_court_neuroscience_may_change_the_legal_system.html.

[28] Briggs, B. (2014, September 24). Guilty conscience? Brain wave breakthrough may reveal crooks. *NBC News*. Retrieved February 2, 2015 from http://www.nbcnews.com/science/science-news/guilty-conscience-brain-wave-breakthrough-may-reveal-crooks-n210191.

[29] Ibid.

[30] Letter from President Barack Obama to Dr. Amy Gutmann, Chair, Presidential Commission for the Study of Bioethical Issues (PCSBI). (2013, July 1). Retrieved February 2, 2015 from http://bioethics.gov/sites/default/files/news/Charge%20from%20President%20Obama.pdf.

[31] Owen, J., Director, MacArthur Foundation Research Network on Law and Neuroscience. (2014, March 31). Comments submitted to the Presidential Commission for the Study of Bioethical Issues (PCSBI).

[32] Topping N. (1952). The United States Public Health Service's Clinical Center for Medical Research. *JAMA*, 150(6), 541-545, p. 545.

[33] National Institute of Neurological Disorders and Stroke (NINDS). (2014). NINDS Overview. Retrieved February 2, 2015 from http://www.ninds.nih.gov/about_ninds/ninds_overview.htm; BrainFacts.org. (2014). Global Burden of Neurological and Mental Disorders [Webpage]. Retrieved February 17, 2015 from http://www.brainfacts.org/policymakers/global-burden-of-neurological-and-mental-disorders/.

[34] World Health Organization (WHO). (2006). *Neurological Disorders: Public Health Challenges*. Geneva, Switzerland: WHO. Retrieved February 2, 2015 from http://www.who.int/mental_health/neurology/neurodiso/en/.

[35] National Institute of Neurological Disorders and Stroke (NINDS). (2005). Neurological Diagnostic Tests and Procedures. Retrieved February 3, 2015 from http://www.ninds.nih.gov/disorders/misc/diagnostic_tests.htm; McKhann, G.M. (2002). Neurology: Then, now and in the future. *Archives of Neurology*, 59(5), 1369-1373; Tyler, K., et al. (2003). Part 2. History of 20th century neurology: Decade by decade. *Annals of Neurology*, 53(4), S27-S45.

[36] McKhann, G.M., *supra* note 35; Tyler, K., et al., *supra* note 35.

[37] Jock Murray, T. (2004). *Multiple Sclerosis: The History of a Disease*. New York, NY: Demos Health; Lublin, F. (2005). History of modern multiple sclerosis therapy. *Journal of Neurology*, 252(S3), S3-S9; Northwestern Medicine. (2015). Multiple sclerosis (MS). Retrieved February 6, 2015 from http://neurology.nm.org/multiple-sclerosis.html.

[38] National Multiple Sclerosis Society. (2014). Medications [Webpage]. Retrieved February 3, 2015 from http://www.nationalmssociety.org/Treating-MS/Medications; National Multiple Sclerosis Society. (2014). The MS Disease-Modifying Medications: General Information [Webpage]. Retrieved February 3, 2015 from http://www.nationalmssociety.org/NationalMSSociety/media/MSNationalFiles/Brochures/Brochure-The-MS-Disease-Modifying-Medications.pdf.

[39] Centers for Disease Control and Prevention (CDC). (2014). Traumatic brain injury in the United States: Fact sheet [Webpage]. Retrieved February 3, 2015 from http://www.cdc.gov/traumaticbraininjury/get_the_facts.html; WHO, *supra* note 34, Section 3.10 Traumatic brain injuries; CDC. (n.d.). Traumatic Brain Injury in the United States: A Report to Congress [Webpage]. Retrieved February 3, 2015 from http://www.cdc.gov/traumaticbraininjury/tbi_report_to_congress.html.

[40] Gravel, J., et al. (2013). Interventions provided in the acute phase for mild traumatic brain injury: A systematic review. *Systematic Reviews*, 2(63), 1-12.

[41] Glannon, W. (2008). Neurostimulation and the minimally conscious state. *Bioethics*, 22(6), 337-345; Le, T.H., and A.D. Gean. (2009). Neuroimaging of traumatic brain injury. *The Mount Sinai Journal of Medicine*, 76(2), 145-162.

[42] Fernandez-Espejo, D., and A.M. Owen. (2013). Detecting awareness after severe brain injury. *Nature Reviews Neuroscience*, 14, 801-809.

[43] World Health Organization (WHO). (2012). Epilepsy [Webpage]. Retrieved February 3, 2015 from http://www.who.int/mediacentre/factsheets/fs999/en/; Centers for Disease Control and Prevention (CDC). (2013). Epilepsy: Frequently Asked Questions [Webpage]. Retrieved February 4, 2105 from http://www.cdc.gov/epilepsy/basics/faqs.htm; National Institute of Neurological Disorders and Stroke (NINDS). (2014). Seizures and Epilepsy: Hope Through Research. Retrieved February 4, 2015 from http://www.ninds.nih.gov/disorders/epilepsy/detail_epilepsy.htm.

[44] Kuzniecky, R.I. (2005). Neuroimaging of epilepsy: Therapeutic implications. *NeuroRx*, 2(2), 384-393.

[45] Ross, C.A., and M.A. Poirier. (2004). Protein aggregation and neurodegenerative disease. *Nature Medicine*, 10(7), S10-S17.

[46] Soto, C. (2003). Unfolding the role of protein misfolding in neurodegenerative diseases. *Nature Reviews Neuroscience*, 4(1), 49-60.

[47] Lozano, A.M., et al. (2002). Deep brain stimulation for Parkinson's disease: Disrupting the disruption. *Lancet Neurology*, 1(4), 225-231.

[48] In 2014, the Lasker Prize, a prestigious prize for scientific accomplishment in the United States, was awarded to neurologist Mahlon DeLong and neurosurgeon Alim Benabid for their contributions to the development of DBS. Lasker Foundation. (2014). Lasker-DeBakey: Clinical Medical Research Award [Webpage]. Retrieved February 19, 2015 from http://www.laskerfoundation.org/awards/2014_c_description.htm.

[49] White House. (2014, September 30). BRAIN Initiative [Webpage]. Retrieved February 3, 2015 from http://www.whitehouse.gov/share/brain-initiative.

[50] Letter from President Barack Obama, to Dr. Amy Gutmann, *supra* note 30; Obama, B. (2013); Remarks by the President on the BRAIN Initiative and American Innovation. Retrieved February 3, 2015 from http://www.whitehouse.gov/the-press-office/2013/04/02/remarks-president-brain-initiative-and-american-innovation.

[51] PCSBI, *supra* note 1.

[52] Allhoff, F., Lin, P., and J. Steinberg. (2009). Ethics of human enhancement: An executive summary. Retrieved February 3, 2015 from http://digitalcommons.calpoly.edu/cgi/viewcontent.cgi?article=1002&context=phil_fac.

[53] Harris, J. (2007). *Enhancing Evolution: The Ethical Case for Making Better People*. Princeton, NJ: Princeton University Press; Sandberg, A., and J. Savulescu. (2011). The social and economic impacts of cognitive enhancement. In J. Savulescu, R. ter Meulen, and G. Kahane (Eds.). *Enhancing Human Capacities*. Malden, MA: Wiley-Blackwell; Savulescu, J. (2006). Justice, fairness, and enhancement. *Annals of the New York Academy of Science*, 1093, 321-338; Buchanan, A. (2011). *Better than Human: The Promise and Perils of Enhancing Ourselves (Philosophy in Action)*. New York, NY: Oxford University Press.

[54] Sandel, M. J. (2002). What's wrong with enhancement. Retrieved February 3, 2015 from https://bioethicsarchive.georgetown.edu/pcbe/background/sandelpaper.html; President's Council on Bioethics. (2003). *Beyond Therapy: Biotechnology and the Pursuit of Happiness*. New York, NY: Regan Books.

[55] Franke, A.G., et al. (2012). Attitudes toward cognitive enhancement in users and nonusers of stimulants for cognitive enhancement: A pilot study. *AJOB Primary Research*, 3(1), 48-57.

[56] Evers, K. (2007). Perspectives on memory manipulation: Using beta-blockers to cure post-traumatic stress disorder. *Cambridge Quarterly of Healthcare Ethics*, 16(2), 138-146.

[57] Pisapia, J.M., et al. (2013). Ethical considerations in deep brain stimulation for the treatment of addiction and overeating associated with obesity. *AJOB Neuroscience*, 4(2), 35-46.

[58] Andrade, C., et al. (2015). Antipsychotic augmentation with modafinil or armodafinil for negative symptoms of schizophrenia: Systematic review and meta-analysis of randomized controlled trials. *Journal of Psychiatric Research*, 60, 14-21; Cakic, V. (2009). Smart drugs for cognitive enhancement: Ethical and pragmatic considerations in the era of cosmetic neurology. *Journal of Medical Ethics*, 35(10), 611-615.

[59] Sahakian, B.J., and S. Morein-Zamir. (2011). Neuroethical issues in cognitive enhancement. *Journal of Psychopharmacology*, 25(2), 197-204.

[60] Bostrom, N., and A. Sandberg. (2009). Cognitive enhancement: Methods, ethics, regulatory challenges. *Science and Engineering Ethics*, 15(3), 311-341.

[61] British Medical Association. (2007). Boosting your brainpower: Ethical aspects of cognitive enhancements. A discussion paper from the British Medical Association. London: British Medical Association. Retrieved February 4, 2015 from http://enhancingresponsibility.com/wp-content/uploads/2014/01/Boosting_brainpower_tcm41-147266.pdf.

[62] Colombo, J., et al., *supra* note 2.

[63] Walsh, R. (2011). Lifestyle and mental health. *American Psychologist*, 66(7), 579-592; Lucke, J., and B. Partridge, *supra* note 3; Taylor, M.P., et al., *supra* note 2.

[64] Presidential Commission for the Study of Bioethical Issues (PCSBI). (2013, December). *Anticipate and Communicate: Ethical Management of Incidental and Secondary Findings in the Clinical, Research, and Direct-to-Consumer Contexts*. Washington, DC: PCSBI, p. 51.

[65] Hardin, D.S. (2008, December). Treatment of short stature and growth hormone deficiency in children with somatotropin (rDNA origin). *Biologics*, 2(4), 655-661.

[66] Ambler, G.R., Fairchild, J., and D.J.C. Wilkinson. (2012). Debate: Idiopathic short stature should be treated with growth hormone. *Journal of Paediatrics and Child Health*, doi:10.1111/j.1440 1754.2012.02465.x.

[67] Buchanan, A., et al. (2001). *From Chance to Choice: Genetics and Justice*. New York, NY: Cambridge University Press.

[68] Levy, N. (2007). *Neuroethics: Challenges for the 21st Century*. New York, NY: Cambridge University Press, pp. 37-44.

[69] Sandel, M.J. (2009). The Case Against Perfection: What's Wrong with Designer Children, Bionic Athletes, and Genetic Engineering, In N. Bostrom and J. Savulescu. (Eds.). *Human Enhancement* (pp. 71-89). New York, NY: Oxford University Press. For a response to Sandel, see Kamm, F.M. (2005). Is there a problem with enhancement? *American Journal of Bioethics*, 5(3), 5-14.

[70] Maslen, H., Faulmüller, N., and J. Savulescu. (2014). Pharmacological cognitive enhancement—How neuroscientific research could advance ethical debate. *Frontiers in Systems Neuroscience*, 8(Article 107), 1-12.

[71] Buchanan, A., *supra* note 5.

[72] WHO, *supra* note 34.

[73] Ibid.

[74] Alzheimer's Association. (2014). 2014 Alzheimer's Disease Facts and Figures, p. 16. Retrieved February 3, 2015 from http://www.alz.org/downloads/Facts_Figures_2014.pdf. Cost includes 2014 cost estimates of Alzheimer's disease and other dementias.

[75] National Parkinson Foundation. (n.d.). Parkinson's Disease Overview [Webpage]. Retrieved February 3, 2015 from http://www.parkinson.org/parkinson-s-disease.aspx.

[76] Epilepsy Foundation. (2014). Epilepsy Statistics [Webpage]. Retrieved February 3, 2015 from http://www.epilepsy.com/learn/epilepsy-statistics.

[77] Centers for Disease Control and Prevention (CDC). (2013). Data & Statistics for Cerebral Palsy [Webpage]. Retrieved February 3, 2015 from http://www.cdc.gov/ncbddd/cp/data.html.

[78] Centers for Disease Control and Prevention (CDC). (2014). Prevalence of Autism Spectrum Disorder among children aged 8 years—Autism and Developmental Disabilities Monitoring Network, 11 sites, United States,

2010. *Morbidity and Mortality Weekly Report – Surveillance Summaries*, 63(SS02), 1-21. Retrieved February 3, 2015 from http://www.cdc.gov/mmwr/preview/mmwrhtml/ss6302a1.htm?s_cid=ss6302a1_w; CDC. (2014). 10 Things to Know about New Autism Data [Webpage]. Retrieved February 3, 2015 from http://www.cdc.gov/features/dsautismdata/.

79 Orton, S.M. (2006). Sex ratio of multiple sclerosis in Canada: A longitudinal study. *The Lancet Neurology*, 5(11), 932-936.

80 WHO, *supra* note 34; Belson, K. (2014, September 12). Brain trauma to affect one in three players, N.F.L. agrees. *New York Times*. Retrieved February 3, 2015 from http://www.nytimes.com/2014/09/13/sports/football/actuarial-reports-in-nfl-concussion-deal-are-released.html.

81 Brain Trauma Foundation. (n.d.). TBI Statistics [Webpage]. Retrieved February 3, 2015 from https://www.braintrauma.org/tbi-faqs/tbi-statistics/.

82 Brain Injury Association of America. (2012). Brain Injury Statistics. Retrieved February 3, 2015 from http://www.biausa.org/LiteratureRetrieve.aspx?ID=104992.

83 National Alliance on Mental Illness. (2013). Mental Illness Facts and Numbers. Retrieved February 4, 2015 from http://www.nami.org/factsheets/mentalillness_factsheet.pdf.

84 Ibid.

85 Mark, T.L., et al. (2011). Changes in US spending on mental health and substance abuse treatment, 1986-2005, and implications for policy. *Health Affairs*, 30(2), 284-292; Insel, T.R. (2008). Assessing the economic costs of serious mental illness. *The American Journal of Psychiatry*, 165(6), 663-665.

86 Bostrom, N. (2014). *Superintelligence: Paths, Dangers, Strategies*. New York, NY: Oxford University Press.

87 Sandberg, A., and N. Bostrom. (2007). Converging cognitive enhancements. *Annals of the New York Academy of Sciences*, 1093, 201-227; Farah, M.J., et al. (2004). Neurocognitive enhancement: What can we do and what should we do? *Nature Reviews Neuroscience*, 5(5), 421-425; Wolpe, P.R. (2002). Treatment, enhancement, and the ethics of neurotherapeutics. *Brain and Cognition*, 50(3), 387-397; Cakic, V., *supra* note 58.

88 Lee, M.A., and B. Shlain. (2007). *Acid Dreams: The Complete Social History of LSD*. New York, NY: Grove Press.

89 Ritalin® is produced by Novartis Pharmaceuticals Corp., East Hanover, New Jersey. Cakic, V., *supra* note 58; Outram, S.M. (2010). The use of methylphenidate among students: The future of enhancement? *Journal of Medical Ethics*, 36(4), 198-202; Sahakian, B.J., and S. Morein-Zamir, *supra* note 59.

90 Brantigan, C.O., Brantigan, T.A., and N. Joseph. (1982). Effect of beta blockade and beta stimulation on stage fright. *American Journal of Medicine*, 72(1), 88-94; Henry, M., Fisher, J.R., and S.J. Younger. (2007). Propranolol and the prevention of post-traumatic stress disorder: Is it wrong to erase the "sting" of bad memories? *American Journal of Bioethics*, 7(9), 12-20; Tindall, B. (2004, October 21). Mixed reviews for a stage-fright remedy. *New York Times*. Retrieved February 4, 2015 from http://www.nytimes.com/2004/10/20/arts/20iht-jitters.html?pagewanted=all.

91 Liao, S.M., and A. Sandberg. (2008). The normativity of memory modification. *Neuroethics*, 1(2), 85-99.

92 Chatterjee, A. (2009). Is it acceptable for people to take methylphenidate to enhance performance. No. *BMJ*, 338, b1956; Fenton, E. (2010). The perils of failing to enhance: A response to Persson and Savulescu. *Journal of Medical Ethics*, 36(3), 148-151; Goodman, R. (2014). Humility pills: Building an ethics of cognitive enhancement. *Journal of Medicine and Philosophy*, 39(3), 258-278; Harris, J. (2009). Is it acceptable for people to take methylphenidate to enhance performance? Yes. *BMJ*, 338, b1955.

93 Adderall® is produced by Shire US Inc., Wayne, Pennsylvania. Substance Abuse and Mental Health Services Administration (SAMHSA). (2009). Nonmedical use of Adderall among full-time college students. Retrieved February 4, 2015 from http://www.samhsa.gov/data/2k9/adderall/adderall.htm.

94 Teter, C.J., et al. (2006). Illicit use of specific prescription stimulants among college students: Prevalence, motives, and routes of administration. *Pharmacotherapy*, 26(10), 1501-1510.

95 Maher, B. (2008). Poll results: Look who's doping. *Nature*, 452(7188), 674-675.

96 Farah, M.J., et al., *supra* note 4, p. 99; Morein-Zamir, S., and B.J. Sahakian. (2011). Pharmaceutical cognitive enhancement. In J., Illes and B.J. Sahakian (Eds.). *The Oxford Handbook of Neuroethics* (pp. 229-244). New York, NY: Oxford University Press.

97 Morein-Zamir, S., and B.J. Sahakian, *supra* note 96; Partridge, B.J., et al. (2011). Smart drugs "as common as coffee": Media hype about neuroenhancement. *PLoS One*, 6(11), e28416; Singh, I., Bard, I., and J. Jackson. (2014). Robust resilience and substantial interest: A survey of pharmacological cognitive enhancement among university students in the UK and Ireland. *PLoS One*, 9(10), e105969.

98 Farah, M.J., et al., *supra* note 4, p. 99.

99 Singh, I., Bard, I., and J. Jackson, *supra* note 97.

100 Lucke, J., and B. Partridge, *supra* note 3.

101 Ibid; Farah, M.J., et al., *supra* note 4.

102 Outram, S., *supra* note 89.

103 Cakic, V., *supra* note 58; Maslen, H., et al. (2014). The regulation of cognitive enhancement devices: Extending the medical model. *Journal of Law and the Biosciences*, 1(1), 68-93.

104 Farah, M.J., et al., *supra* note 4.

[105] Repantis, D. et al. (2010). Modafinil and methylphenidate for neuroenhancement in healthy individuals: A systematic review. *Pharmacological Research*, 62, 187-206.
[106] Farah, M.J., et al., *supra* note 4, p. 98.
[107] Morein-Zamir, S., and B.J. Sahakian, *supra* note 96; Graf, W., et al., *supra* note 7.
[108] Chai, G., et al. (2012). Trends of outpatient prescription drug utilization in US children, 2002-2010. *Pediatrics*, 130, 23-31; Graf, W., et al., *supra* note 7; Graf, W.D., Miller, G., and S.K. Nagel. (2014). Addressing the problem of ADHD medication as neuroenhancements. *Expert Reviews of Neurotherapeutics*, 14(5), 569-581.
[109] Johnston, L.D., et al. (2014). *Monitoring the Future National Survey Results on Drug Use, 1975-2013. Volume I: Secondary School Students*. Ann Arbor, MI: Institute for Social Research, University of Michigan and National Institute on Drug Abuse, p. 609. Retrieved February 4, 2015 from http://www.monitoringthefuture.org/pubs/monographs/mtf-vol1_2013.pdf.
[110] Martinez-Raga, J., et al. (2013). Risk of serious cardiovascular problems with medications for attention-deficit hyperactivity disorder. *CNS Drugs*, 27(1), 15-30; Cooper, W.O., et al. (2011). ADHD drugs and serious cardiovascular events in children and young adults. *The New England Journal of Medicine*, 365(20), 1896-1904.
[111] Singh, I., and K.J. Kelleher. (2010). Neuroenhancement in young people: Proposal for research, policy, and clinical management. *AJOB Neuroscience*, 1(1), 3-16; Srisurapanont, M., Jarusuraisin, N., and P. Kittirattanapaiboon. (2001). Treatment for amphetamine dependence and abuse. *The Cochrane Database of Systematic Reviews*, 4, Art. No.: CD003022. DOI: 10.1002/14651858.CD003022.
[112] Kim, Y., et al. (2009). Methylphenidate-induced dendritic spine formation and ΔFosB expression in nucleus accumbens. *Proceedings of the National Academy of Sciences*, 106(8), 2915-2920.
[113] Graf, W., et al., *supra* note 7.
[114] Ibid; Gaucher, N., Payot, M., and E. Racine. (2013). Cognitive enhancement in children and adolescents: Is it in their best interests? *Acta Paediatrica*, 102(12), 1118-1124; Marco, E.M., et al. (2011). Neurobehavioral adaptations to methylphenidate: The issue of early adolescent exposure. *Neuroscience and Biobehavioral Reviews*, 35(8), 1722-1739.
[115] Maslen, H., Faulmüller, N., and J. Savulescu, *supra* note 70.
[116] Allhoff, F., et al. (2009). Ethics of Human Enhancement: 25 Questions & Answers. Retrieved February 4, 2015 from http://www.humanenhance.com/NSF_report.pdf, p. 6.
[117] Savulescu, J., *supra* note 53.
[118] Clarke, R., et al. (2007). Low vitamin B-12 status and risk of cognitive decline in older adults. *American Journal of Clinical Nutrition*, 86(5), 1384-1391.
[119] Klonopin® is produced by Genentech, South San Francisco, California.
[120] Nardi, A.E., et al. (2013). Clonazepam for the treatment of panic disorder. *Current Drug Targets*, 14(3), 353-364; Freire, R.C., et al. (2014). Current pharmacological interventions in panic disorder. *CNS & Neurological Disorders Drug Targets*, 13(6), 1057-1065.
[121] Buchanan, A., *supra* note 5; Farah, M.J., et al., *supra* note 87; Caplan, A.L. (2003). Is better best? *Scientific American*, September 2003, 104-105.
[122] Enriquez-Geppert, S., Huster, R.J., and C.S. Herrmann. (2013). Boosting brain functions: Improving executive functions with behavioral training, neurostimulation, and neurofeedback. *International Journal of Psychophysiology*, 88(1), 1-16.
[123] Bostrom, N., and A. Sandberg, *supra* note 60; British Medical Association, *supra* note 61.
[124] Bostrom, N., and R. Roache. (2011). Smart Policy: Cognitive Enhancement and the Public Interest. In J., Savulescu, R., ter Meulen, and G. Kahane (Eds.). *Enhancing Human Capacities* (pp. 138-149). Malden, MA: Wiley-Blackwell.
[125] Buchanan, A., *supra* note 5.
[126] De Jongh, R., et al. (2008). Botox for the brain: enhancement of cognition, mood and pro-social behavior and blunting of unwanted memories. *Neuroscience and Behavioral Reviews*, 32, 760-776; Housden, C.R., Morein-Zamir, and B.J. Sahakian. (2011). Cognitive enhancing drugs: Neuroscience and Society. In J., Savulescu, R., ter Meulen, and G. Kahane (Eds.). *Enhancing Human Capacities* (pp. 113-126). Malden, MA: Wiley-Blackwell.
[127] Savulescu, J., *supra* note 53.
[128] Caplan, A.L., *supra* note 121.
[129] Johnson, J. (2009). A dark history: Memories of lobotomy in a new era of psychosurgery. *Medicine Studies*, 1(4), 367-378; Feldman, R.P., and J.T. Goodrich. (2001). Psychosurgery: A historical overview. *Neurosurgery*, 48(3), 347-659.
[130] President's Council on Bioethics, *supra* note 54.
[131] Sandel, M.J., *supra* note 54; Habermas, J. (2003). *The Future of Human Nature*. Malden, MA: Polity Press; Fukuyama, F. (2003). *Our Posthuman Future: Consequences of the Biotechnology Revolution*. New York, NY: Picador Press.
[132] President's Council on Bioethics, *supra* note 54.

[133] Ibid; Elliot, C. (2003). *Better Than Well: American Medicine Meets the American Dream*. New York, NY: W.W. Norton & Company.

[134] The President's Council on Bioethics also argued that people with such memories are essential to provide testimony of atrocities to the rest of society. President's Council on Bioethics, *supra* note 54.

[135] Ibid; Elliot, C., *supra* note 133 p. 212.

[136] Zak, P.J., Stanton, A.A., and S. Ahmadi. (2007). Oxytocin increases generosity in humans. *PLoS ONE*, 2(11), e1128; Crockett, M., et al. (2010). Serotonin selectively influences moral judgment and behavior through effects on harm aversion. *Proceedings of the National Academy of Sciences*, 107(40), 17433-17438.

[137] Wiseman, H. (2014). SSRIs as moral enhancement interventions: A practical dead end. *AJOB Neuroscience*, 5(3), 21-30.

[138] Persson, I., and J. Savulescu. (2008). The perils of cognitive enhancement and the urgent imperative to enhance the moral character of humanity. *Journal of Applied of Philosophy*, 25(3), 162-177; Douglas, T. (2008). Moral enhancement. *Journal of Applied Philosophy*, 25(3), 228-245.

[139] Presidential Commission for the Study of Bioethical Issues (PCSBI). (2010, December). *New Directions: The Ethics of Synthetic Biology and Emerging Technologies*. Washington, DC: PCSBI, pp. 155-157.

[140] Caulfield, T. (2004). Biotechnology and the popular press: Hype and the selling of science. *Trends in Biotechnology*, 22(7), 337-339; Caulfield, T., and C. Condit. (2012). Science and the sources of hype. *Public Health Genomics*, 15(3-4), 209-217; Master, Z., and D.B. Resnick. (2013). Hype and public trust in science. *Science and Engineering Ethics*, 19(2), 321-335; Rusconi, E., and T. Mitchener-Nissen, *supra* note 10.

[141] Ibid.

[142] Similar efforts are already underway. BrainFacts.org, sponsored by The Kavli Foundation, The Gatsby Foundation, and the Society for Neuroscience, provides resources designed to educate stakeholders about basic neuroscience principles and the societal implications of neuroscience research and its applications. The information provided on BrainFacts.org is fact-checked, updated, and reviewed for accuracy by members of the Society for Neuroscience or other partners approved by the BrainFacts.org editorial board. See The Kavli Foundation, The Gatsby Foundation, and the Society for Neuroscience. (n.d.). BrainFacts.org. Retrieved February 4, 2015 from http://www.brainfacts.org.

[143] See generally, PCSBI, *supra* note 139.

[144] Ibid.

[145] Ferrie, J.E., et al, *supra* note 2; Colombo, J., et al, *supra* note 2; Taylor, M.P., et al., *supra* note 2; Hu, J., *supra* note 2.

[146] Lucke, J., and B. Partridge, *supra* note 3.

[147] See, e.g., Brinch, C.N., and T.A. Galloway. (2012). Schooling in adolescence raises IQ scores. *Proceedings of the National Academy of Sciences*, 109(2), 425-430; Diekelmann, S. (2014). Sleep for cognitive enhancement. *Frontiers in Systems Neuroscience*, 8(46), 1-12.

[148] Hochberg, L.R., et al. (2012). Reach and grasp by people with tetraplegia using a neurally controlled robotic arm. *Nature*, 485(7398), 372-375.

[149] See, e.g., Jotterand, F., and J. Giordano. (2011). Transcranial magnetic stimulation, deep brain stimulation and personal identity: Ethical questions, and neuroethical approaches for medical practice. *International Review of Psychiatry*, 23(5), 476-485; Cabrera, L.Y., Evans, E.L., and R.H. Hamilton. (2014). Ethics of the electrified mind: Defining issues and perspectives on the principled use of brain stimulation in medical research and clinical care. *Brain Topography*, 27(1), 33-45; Depression and Bipolar Support Alliance. (2009). Treatment Technologies for Mood Disorders. Retrieved February 2, 2015 from http://www.dbsalliance.org/pdfs/EmrgTechsBro09.FINAL.pdf; National Institute of Mental Health. (n.d.). Brain Stimulation Therapies. Retrieved February 2, 2015 from http://www.nimh.nih.gov/health/topics/brain-stimulation-therapies/brain-stimulation-therapies.shtml.

[150] Ragan, C.I., Bard, I., and I. Singh, *supra* note 4; Smith, M.E., and M.J. Farah, *supra* note 4; Ilieva, I., Boland, J., and M.J. Farah, *supra* note 4.

[151] Lucke, J.C., et al. (2011). Deflating the neuroenhancement bubble. *AJOB Neuroscience*, 2(4), 38-43; Farah, M.J., et al. (2014). Cognitive enhancement. *WIREs Cognitive Science*, 5(1), 95-103; Morein-Zamir, S., and B.J. Sahakian, *supra* note 96.

[152] Reddit. (2014). tDCS [Web forum]. Retrieved February 2, 2015 from http://www.reddit.com/r/tDCS/; Maslen, H., et al. (2014). The regulation of cognitive enhancement devices: Extending the medical model. *Journal of Law and the Biosciences*, 68-93; Farah, M.J., et al., *supra* note 4.

[153] Racine, E., and C. Forlini. (2010). Cognitive enhancement, lifestyle choice or misuse of prescription drugs? Ethics blind spots in current debates. *Neuroethics*, 3(1), 1-4.

[154] Buchanan, A., *supra* note 5; Greely, H., et al., *supra* note 5.

[155] Morein-Zamir, S., and B.J. Sahakian, *supra* note 96; Maslen, H., Faulmüller, N., and J. Savulescu, *supra* note 70.

[156] Maslen, H., Faulmüller, N., and J. Savulescu, *supra* note 70; Bostrom, N., and A. Sandberg, *supra* note 60.

[157] Hotze, T.D., et al., *supra* note 6.

[158] Larriviere, D., et al., *supra* note 6.

[159] Graf, W., et al., *supra* note 7.

[160] Greely, H., et al., *supra* note 5; Franke, A.G., et al., *supra* note 55.

[161] Reiner, P.B. (2010). Distinguishing between restoration and enhancement in neuropharmacology. *Virtual Mentor*, 12(11), 885-888.

[162] Greely, H., et al., *supra* note 5.

[163] Graf, W., et al., *supra* note 7.

[164] Ibid.

[165] Ibid.

[166] National Commission for the Protection of Human Subjects of Biomedical and Behavioral Research (National Commission). (1978). *The Belmont Report: Ethical Principles and Guidelines for the Protection of Human Subjects of Research* (DHEW Publication No. OS 78-0012). Washington, DC: Department of Health, Education, and Welfare. Retrieved February 2, 2015 from http://www.hhs.gov/ohrp/humansubjects/guidance/belmont.html.

[167] Emanuel, E.J., Wendler, D., and C. Grady. (2000). What makes clinical research ethical? *JAMA*, 283(20), 2701-2711.

[168] Appelbaum, P.S., and T. Grisso. (1995). The MacArthur Treatment Competence Study. I: Mental illness and competence to consent to treatment. *Law and Human Behavior*, 19(2), 105-126.

[169] Labuzetta, J.N., Burnstein, R., and J. Pickard, *supra* note 8; Palmer, B.W., et al. (2013). Changes in capacity to consent over time in patients involved in psychiatric research. *British Journal of Psychiatry*, 202, 454-458; Appelbaum, P.S., *supra* note 8.

[170] Dybedal, G.S., et al. (2014). Cognitive side-effects of electroconvulsive therapy in elderly depressed patients. *The Clinical Neuropsychologist*, 28(7), 1071-1090.

[171] Muthappan, P., Forster, H., and D. Wendler. (2005). Research advance directives: Protection or obstacle? *American Journal of Psychiatry*, 162(12), 2389-2391; Stocking, C.B., et al. (2006). Speaking of research advance directives: Planning for future research participation. *Neurology*, 66(9), 1361-1366.

[172] Palmer, B.W., and G.N. Savla. (2007). The association of specific neuropsychological deficits with capacity to consent to research or treatment. *Journal of the International Neuropsychological Society*, 13(6), 1047-1059.

[173] National Commission for the Protection of Human Subjects of Biomedical and Behavioral Research (National Commission). (1978). *Research Involving Those Institutionalized as Mentally Infirm*. Washington, DC: Department of Health, Education, and Welfare. Retrieved February 2, 2015 from http://videocast.nih.gov/pdf/ohrp_research_mentally_infirm.pdf; National Commission for the Protection of Human Subjects of Biomedical and Behavioral Research (National Commission). (1977). *Psychosurgery*. Washington, DC: Department of Health, Education, and Welfare. Retrieved February 2, 2015 from http://videocast.nih.gov/pdf/ohrp_psychosurgery.pdf; Beecher, H.K. (1966). Ethics and clinical research. *New England Journal of Medicine*, 274(24), 1354-1360.

[174] Beauchamp, T.L., and J.F. Childress. (2013). Respect for Autonomy. In *Principles of Biomedical Ethics*. (pp. 101-149). Seventh Edition. New York, NY: Oxford University Press.

[175] Lebacqz, K. (2005). We sure are older but are we wiser? In *Belmont Revisited: Ethical Principles for Research with Human Subjects*, (pp. 99-111). Childress, J.F., Meslin, E.M., and H.T. Shapiro. (Eds.). Washington, DC: Georgetown University Press.

[176] Black, B.S., et al. (2010). Seeking assent and respecting dissent in dementia research. *American Journal of Geriatric Psychiatry*, 18(1), 77-85.

[177] Brudney, D. (2009). Choosing for another: Beyond autonomy and best interests. *Hastings Center Report*, 39(2), 31-37; Jaworska, A. (1999). Respecting the margins of agency: Alzheimer's patients and their capacity to value. *Philosophy & Public Affairs*, 28(2), 105-138; Lindemann, H. (2014). *Holding and Letting Go: The Social Practice of Personal Identifies*. New York, NY: Oxford University Press; Meninger, H.P. (2001). Authenticity in community: Theory and practice of an inclusive anthropology in care for persons with intellectual disabilities. *Journal of Religion, Disability, & Health*, 5(2-3), 13-28.

[178] Berghmans, R.L.P. (1998). Advance directives for non-therapeutic dementia research: Some ethical and policy considerations. *Journal of Medical Ethics*, 24, 32-37.

[179] Berg, J.W., and P.S. Appelbaum. (1999). Subjects' Capacity to Consent to Neurobiological Research. In H.A. Pincus, J.A. Lieberman, and S. Ferris (Eds.). *Ethics in Psychiatric Research: A Resource Manual for Human Subjects Protection* (pp. 81-106). Arlington, VA: American Psychiatric Publishing, p. 95.

[180] PCSBI, *supra* note 143, pp. 24-25.

[181] Appelbaum, P.S. (2002). Involving decisionally impaired subjects in research: The need for legislation. *American Journal of Geriatric Psychiatry*, 10(2), 120-124.

[182] In addition to the extended time it would have taken to enroll participants had surrogate consent not been used in the NINDS rt-PA Stroke Trial, those enrolled would likely have differed in consequential ways for the generalizability of the results. For example, "subjects enrolled by surrogate consent were older and had more severe strokes than persons enrolled by their own consent. Results from a study using only self-consent would have left clinicians caring for older and sicker patients struggling to determine whether rt-PA was appropriate for this group." Flaherty, M.L., et al. (2008). How important is surrogate consent for stroke research? *Neurology*, 71(20), 1566-1571, p. 1569.

[183] National Commission, *supra* note 166.
[184] See generally, Flaherty, M.L., et al., *supra* note 182.
[185] National Commission, *supra* note 166.
[186] Presidential Commission for the Study of Bioethical Issues (PCSBI). (2013, March). *Safeguarding Children: Pediatric Medical Countermeasure Research*. Washington, DC: PCSBI.
[187] *Protection of Human Subjects, HHS*. 45 C.F.R. Part 46, Subpart B.
[188] *Protection of Human Subjects, HHS*. 45 C.F.R. Part 46, Subpart A.
[189] Levine, C., et al. (2004). The limitations of "vulnerability" as a protection for human research participants. *American Journal of Bioethics*, 4(3), 44-49.
[190] Corrigan, P., Distinguished Professor of Psychology, Illinois Institute of Technology. (2014). Beating the Stigma of Mental Illness. Presentation to the Presidential Commission for the Study of Bioethical Issues, June 9. Retrieved February 2, 2015 from http://bioethics.gov/node/3777; Schonfeld, T. (2015). Research Involving Vulnerable Populations: A Critical Analysis. In J.D. Arras, E. Fenton, and R. Kukla (Eds.). *The Routledge Companion to Bioethics* (pp. 238-248). New York, NY: Routledge.
[191] Levine, C., et al., *supra* note 189.
[192] Corrigan, P., *supra* note 190.
[193] Levine, C., et al., *supra* note 189, p. 48.
[194] Feinberg, J. (1988). *Harmless Wrongdoing*. New York, NY: Oxford University Press; Okin, S.M. (1989). *Justice, Gender, and the Family*. New York, NY: Basic Books; Wertheimer, A. (1999). *Exploitation*. Princeton, NJ: Princeton University Press.
[195] Wertheimer, A. (2010). *Rethinking the Ethics of Clinical Research: Widening the Lens*. New York, NY: Oxford University Press.
[196] Shore, D., and S.E. Hyman. (1999). An NIMH commentary on the NBAC report. *Biological Psychiatry*, 46, 1013-1016; Michels, R. (1999). Are research ethics bad for our mental health? *New England Journal of Medicine*, 340(18), 1427-1430; Oldham, J.M., Haimowitz, S., and S.J. Delano. (1999). Protection of persons with mental disorders from research risk: A response to the report of the National Bioethics Advisory Commission. *Archives of General Psychiatry*, 56(8), 688-693.
[197] Mastroianni, A., and J. Kahn. (2001). Swinging on the pendulum: Shifting views of justice in human subjects research. *Hastings Center Report*, 31(3), 21-28.
[198] Ibid.
[199] Kant, I. (1785). Groundwork of The Metaphysics of Morals. In M.J. Gregor (Ed.). *Practical Philosophy: The Cambridge Edition of the Works of Immanuel Kant* (pp. 37-108). (1996). New York, NY: Cambridge University Press; PCSBI, *supra* note 186.
[200] Mental Health Foundation. (2012). Mental Capacity and the Mental Capacity Act 2005: A Literature Review. Retrieved February 3, 2015 from http://www.mentalhealth.org.uk/content/assets/PDF/publications/mca-lit-review.pdf.
[201] Pescosolido, B.A., et al. (2008). Rethinking theoretical approaches to stigma: A Framework Integrating Normative Influences on Stigma (FINIS). *Social Science & Medicine*, 67(3), 431-440.
[202] National Bioethics Advisory Commission (NBAC). (1998). *Research Involving Persons with Mental Disorders That May Affect Decisionmaking Capacity*. Volume 1. Bethesda, MD: NBAC. Retrieved February 3, 2015 from https://bioethicsarchive.georgetown.edu/nbac/capacity/TOC.htm; National Commission, *Research Involving Those Institutionalized as Mentally Infirm*, *supra* note 173, pp. 15-16, 119.
[203] Oldham, J.M., Haimowitz, S., and S.J. Delano, *supra* note 196.
[204] Mental Health Foundation, *supra* note 200.
[205] Corrigan, P., *supra* note 190.
[206] Ibid.
[207] Schomerus, G., et al. (2012). Evolution of public attitudes about mental illness: A systematic review and meta-analysis. *Acta Psychiatrica Scandinavica*, 125(6), 440-452.
[208] Corrigan, P., *supra* note 190.
[209] National Commission, *Psychosurgery*, *supra* note 173.
[210] National Commission, *Research Involving Those Institutionalized as Mentally Infirm*, *supra* note 173, p. 58.
[211] Protection of Human Subjects: Proposed Regulations on Research Involving Those Institutionalized as Mentally Disabled, 43 Fed. Reg. 53,950, 53,956. (proposed Nov. 17, 1978); Tovino, S.A. (2013). A "Common" proposal. *Houston Law Review*, 50(3), 787-854, p. 799.
[212] NBAC, *supra* note 202, pp. 51-67.
[213] Eiseman, E. (2003). *The National Bioethics Advisory Commission: Contributing to Public Policy*. Arlington, VA: RAND. Retrieved February 3, 2015 from http://www.rand.org/content/dam/rand/pubs/monograph_reports/2007/MR1546.pdf.
[214] Eiseman, E., *supra* note 213; HHS Working Group on the NBAC Report. (2001). Analysis and Proposed Actions Regarding the NBAC Report: Research Involving Persons with Mental Disorders that May Affect Decisionmaking Capacity. Washington, DC: Department of Health and Human Services. Retrieved February

3, 2015 from http://aspe.hhs.gov/sp/nbac/nbac.pdf; Oldham, J.M., Haimowitz, S., and S.J. Delano, *supra* note 196.

[215] Ibid.

[216] Eiseman, E., *supra* note 213; National Institute of Neurological Disorders and Stroke (NINDS). (1999). NINDS Advisory Council meeting minutes, February 11-12, 1999. Retrieved February 3, 2015 from http://www.ninds.nih.gov/find_people/nands/council_minutes_february1999.htm.

[217] Eiseman, E., *supra* note 213.

[218] Eiseman, E., *supra* note 213, p. 48; Shore, D., and S.E. Hyman, *supra* note 196, p. 1016.

[219] National Institutes of Health (NIH) Office of Extramural Research. (2009). Research Involving Individuals with Questionable Capacity to Consent: Points to Consider. Retrieved February 3, 2015 from http://grants.nih.gov/grants/policy/questionablecapacity.htm.

[220] Secretary's Advisory Committee on Human Research Protections (SACHRP). (2009). Recommendations from the Subcommittee for the Inclusion of Individuals with Impaired Decision Making in Research (SIIIDR). Retrieved February 3, 2015 from http://www.hhs.gov/ohrp/sachrp/20090715letterattach.pdf.

[221] Ibid.

[222] Tovino, S.A., *supra* note 211, p. 805.

[223] *Protection of Human Subjects, HHS*. 45 C.F.R. §§ 46.109, 46.116. Note, however, that the Common Rule provides for the possibility of a waiver of informed consent in certain circumstances. See *Protection of Human Subjects, HHS*. 45 C.F.R. § 46.116(d).

[224] *Protection of Human Subjects, HHS*. 45 C.F.R. § 46.111.

[225] *Protection of Human Subjects, HHS*. 45 C.F.R. Part 46, Subparts B, C, D.

[226] *Protection of Human Subjects, HHS*. 45 C.F.R. § 46.111.

[227] Office for Human Research Protections (OHRP). (n.d.). Informed Consent FAQs. Retrieved February 3, 2015 from http://www.hhs.gov/ohrp/policy/faq/informed-consent/index.html.

[228] Office for Human Research Protections (OHRP). (n.d.). Informed Consent FAQs: When may a legally authorized representative provide consent on behalf of an adult with diminished decision-making capacity? Retrieved February 17, 2015 from http://www.hhs.gov/ohrp/policy/faq/informed-consent/legally-authorized-representative-provide-consent-adult-diminished-decision-making-capacity.html.

[229] Office for Human Research Protections (OHRP). (n.d.). Informed Consent FAQs: What should be considered in seeking informed consent from individuals with diminished decision-making capacity? Retrieved February 17, 2015 from http://www.hhs.gov/ohrp/policy/faq/informed-consent/seeking-informed-consent-from-individuals-with-diminished-decision-making-capacity.html.

[230] *Protection of Human Subjects, FDA*. 21 C.F.R. § 56.107(a), (f); 21 C.F.R. § 50.20.

[231] U.S. Food and Drug Administration (FDA). (2014). Informed Consent Information Sheet: Guidance for IRBs, Clinical Investigators, and Sponsors. Draft Guidance. Retrieved February 3, 2015 from http://www.fda.gov/RegulatoryInformation/Guidances/ucm404975.htm#impaired.

[232] Ibid.

[233] Flaherty, M.L., et al., *supra* note 182; Sabatino, C.P. (1999). The legal and functional status of the medical proxy: Suggestions for statutory reform. *Journal of Law, Medicine & Ethics*, 27(1), 52-68.

[234] Hoffman, D.E., and J. Schwartz. (1998). Proxy consent to participation of the decisionally impaired in medical research—Maryland's policy initiative. *Journal of Health Care Law & Policy*, 1(1), 123-153, pp. 125-129; see generally, Tovino, S.A., *supra* note 211.

[235] Gong, M.N., et al. (2010). Surrogate consent for research involving adults with impaired decision making: Survey of Institutional Review Board practices. *Critical Care Medicine*, 38(11), 2146-2154.

[236] See generally, Tovino, S.A., *supra* note 211.

[237] Sturman, E.D. (2005). The capacity to consent to treatment and research: A review of standardized assessment tools. *Clinical Psychology Review*, 25(7), 954-974; Wendler, D., and K. Prasad. (2001). Core safeguards for clinical research with adults who are unable to consent. *Annals of Internal Medicine*, 135(7), 514-523.

[238] Dunn, L.B., et al. (2006). Assessing decisional capacity for clinical research or treatment: A review of instruments. *American Journal of Psychiatry*, 163(8), 1323-1334; Fisher, C.B., and M. Oransky. (2008). Informed consent to psychotherapy: Protecting the dignity and respecting the autonomy of patients. *Journal of Clinical Psychology: In Session*, 64(5), 576-588. Structured interviews might help to counter faulty intuitions. For example, clinicians are more likely to question capacity when patients disagree with clinical advice; Sturman, E.D., *supra* note 237.

[239] Dunn, L.B., et al., *supra* note 238.

[240] Appelbaum, P.S., and T. Grisso. (2001). *MacArthur Competence Assessment Tool for Clinical Research (MaCAT-CR)*. Sarasota, FL: Professional Resource Press.

[241] Dunn, L.B., et al., *supra* note 238; Gupta, U.C., and S. Kharawala. (2012). Informed consent in psychiatry clinical research: A conceptual review of issues, challenges, and recommendations. *Perspectives in Clinical Research*, 3(1), 8-15.

[242] Jeste, D.V., et al. (2009). Multimedia consent for research in people with schizophrenia and normal subjects: A randomized trial. *Schizophrenia Bulletin*, 35(4), 719-729.

[243] Ibid; Kaup, A.R., et al. (2011). Decisional capacity to consent to research in schizophrenia: An examination of errors. *IRB*, 33(4), 1-9.

[244] Black, B.S., et al, *supra* note 176. The authors acknowledge that their recommendations might be extended to other domains in clinical neuroscience, including research involving participants who have experienced traumatic brain injury, stroke, or intellectual disability; however, they explicitly caution against using similar standards in a clinical context.

[245] Black, B.S., et al, *supra* note 176.

[246] Ibid; Saughter, S., et al. (2007). Consent and assent to participate in research from people with dementia. *Nursing Ethics*, 14(1), 27-40.

[247] Black, B.S., et al, *supra* note 176.

[248] Ibid.

[249] Weijer, C., et al. (1995). Monitoring clinical research: An obligation unfulfilled. *Canadian Medical Association Journal*, 152(12), 1973-1980.

[250] National Commission, *Research Involving Those Institutionalized as Mentally Infirm*, *supra* note 173.

[251] *Protection of Human Subjects, HHS*. 45 C.F.R. Part 46; *Institutional Review Boards, FDA*. 21 CFR § 56.109.

[252] Silverman, H. (2011). Protecting vulnerable research subjects in critical care trials: Enhancing the informed consent process and recommendations for safeguards. *Annals of Intensive Care*, 1(1), 1-7; Weijer, C., et al, *supra* note 249.

[253] Slaughter, S., et al. (2007). Consent and assent to participate in research from people with dementia. *Nursing Ethics*, 14(1), 27-40.

[254] Rosenstein, D.L., and F.G. Miller. (2003). Ethical considerations in psychopharmacological research involving decisionally impaired subjects. *Psychopharmacology*, 171(1), 92-97.

[255] Although some researchers contend that this articulation conflates autonomy and beneficence, disregard for the wellbeing of those who cannot direct the course of their own lives can also be interpreted as disrespectful treatment of valued members of society. Limiting risks of research thus can be said to follow from both the principles of respect for persons and beneficence. National Commission, *supra* note 166; PCSBI, *supra* note 186; Beauchamp, T.L. (2005). The Origins and Evolution of the Belmont Report. In J.F. Childress, E.M. Meslin, and H.T. Shapiro (Eds.). *Belmont Revisited: Ethical Principles for Research with Human Subjects* (pp. 12-25). Washington, DC: Georgetown University Press, pp. 15-16.

[256] New York State Task Force on Life and the Law. (2014). *Report and Recommendations for Research With Human Subjects Who Lack Consent Capacity*. Retrieved February 3, 2015 from https://www.health.ny.gov/regulations/task_force/docs/report_human_subjects_research.pdf.

[257] National Commission, *Research Involving Those Institutionalized as Mentally Infirm*, *supra* note 173, pp. 120-121; NBAC, *supra* note 202, pp. 51-67; SACHRP, *supra* note 220.

[258] National Commission, *Research Involving Those Institutionalized as Mentally Infirm*, *supra* note 173, p. 120.

[259] *Protection of Human Subjects, HHS*. 45 C.F.R Part 46, Subpart D; *Protection of Human Subjects, FDA*. 21 C.F.R. Part 50, Subpart D.

[260] New York State Task Force on Life and the Law, *supra* note 256.

[261] National Commission, *Research Involving Those Institutionalized as Mentally Infirm*, *supra* note 173.

[262] SACHRP, *supra* note 220; New York State Task Force on Life and the Law, *supra* note 256.

[263] Wendler, D., and K. Prasad, *supra* note 237; Buchanan, A.E., and D.W. Brock. (1990). *Deciding for Others: The Ethics of Surrogate Decision Making*. New York, NY: Cambridge University Press, pp. 112-116.

[264] The substituted judgment standard can be problematic, even in clinical care. Surrogates are often inaccurate in predicting what patients might want, and most patients do not want their surrogates to use a pure substituted judgment standard. See, e.g., Sulmasy, D.P. (2010). Research in Medical Ethics: Scholarship in Substituted Judgment. In J. Sugarman and D.P. Sulmasy (Eds.). *Methods in Medical Ethics*, Second Edition. Washington, DC; Georgetown University Press, pp. 295-314.

[265] Brudney, D., *supra* note 177; Sulmasy, D.P., and L. Snyder. (2010). Substituted interests and best judgments: An integrated model of surrogate decision making. *JAMA*, 304(17), 1946-1947.

[266] Fagerlin, A., and C.E. Schneider. (2004). Enough: The failure of the living will. *Hastings Center Report*, 34(2), 30-42.

[267] Wendler, D., and K. Prasad, *supra* note 237.

[268] NIH. (n.d.). NIH Advance Directive for Health Care and Medical Research Participation, NIH-200 (10-00).

[269] Participants in the Community Engagement and Consent Workshop. (2013). Consent and community engagement in diverse research contexts: Reviewing and developing research and practice. *Journal of Empirical Research on Human Research Ethics*, 8(4), 1-18.

[270] PCSBI, *supra* note 1.

[271] MacQueen, K.M., et al. (2012). *Stakeholder Engagement Toolkit for HIV Prevention Trials*. Research Triangle Park, NC: FHI 360. Retrieved February 4, 2015 from http://www.fhi360.org/sites/default/files/media/documents/Stakeholder_Engagement_Toolkit-FULL_0.pdf; Lemke, A.A., et al. (2010). Community engagement in biobanking: Experiences from the eMERGE Network. *Genomics, Society, and Policy*, 6(3), 35-52; Joint United Nations Programme on HIV/AIDS (UNAIDS). (2011). *Good Participatory Practice:*

Guidelines for Biomedical HIV Prevention Trials. Geneva, Switzerland: UNAIDS. Retrieved February 4, 2015 from http://www.unaids.org/sites/default/files/media_asset/JC1853_GPP_Guidelines_2011_en_0.pdf.

[272] MacQueen, K.M., et al., *supra* note 271; Lemke, A.A., et al., *supra* note 271; PCSBI *supra* note 1; NBAC, *supra* note 202, pp. 53-54; *Protection of Human Subjects, HHS*. 45 C.F.R. § 46.107(a).

[273] NBAC, *supra* note 202, pp. 53-54; SACHRP, *supra* note 220, pp. 6-7.

[274] Ahmed, S.M., and A.G. Palermo. (2010). Community engagement in research: Frameworks for education and peer review. *American Journal of Public Health*, 100(8), 1380-1387; National Human Genome Research Institute (NHGRI). (2014). Community engagement at the NIH. Retrieved February 4, 2015 from http://www.genome.gov/27541346.

[275] WHO, *supra* note 34.

[276] Hindmarch, T., Hotopf, M., and G.S. Owen. (2013). Depression and decision-making capacity for treatment or research: A systematic review. *BMC Medical Ethics*, 14(54), 1-10; Cabrera, L. (2011). They might retain capacities to consent but do they even care? *AJOB Neuroscience*, 2(1), 41-42.

[277] Berney, A., et al. (2002). Effect on mood of subthalamic DBS for Parkinson's disease: A consecutive series of 24 patients. *Neurology*, 59(9), 1427-1429.

[278] Bell, E., et al. (2014). Beyond consent in research: Revisiting vulnerability in deep brain stimulation for psychiatric disorders. *Cambridge Quarterly of Healthcare Ethics*, 23(3), 361-368.

[279] Fisher, C.E., et al. (2012). The ethics of research on deep brain stimulation for depression: Decisional capacity and therapeutic misconception. *Annals of the New York Academy of Sciences*, 1265, 69-79; Christopher, P.P., et al. (2012). Enrolling in deep brain stimulation for depression: Influences on potential subjects' decision making. Depression and Anxiety, 29(2), 139-146; Wertheimer, A., *supra* note 195.

[280] Dunn, L.B., et al., *supra* note 238.

[281] Ibid.

[282] Sturman, E.D., *supra* note 237.

[283] Ganzini, L., et al. (2003). Pitfalls in assessment of decision-making capacity. *Psychosomatics*, 44(3), 237-243.

[284] Buchanan, A., and D.W. Brock. (1986). Deciding for others. *Milbank Quarterly*, 64(2), 17-94; Braude, H., and J. Kimmelman. (2012). The ethics of managing affective and emotional states to improve informed consent: autonomy, comprehension, and voluntariness. *Bioethics*, 26(3), 149-156; Roberts, L.W. (2002). Informed consent and the capacity for voluntarism. *American Journal of Psychiatry*, 159(5), 705-712; Dunn, L.B., et al., *supra* note 238.

[285] Saks, E.R., and D.V. Jeste. (2006). Capacity to consent to or refuse treatment and/or research: Theoretical considerations. *Behavioral Sciences & The Law*, 24(4), 411-429, p. 421; Sturman, E.D., *supra* note 237; Gupta, U.C., and S. Kharawala, *supra* note 241.

[286] Appelbaum, P.S., et al. (2012). Therapeutic misconception in research subjects: Development and validation of a measure. *Clinical Trials*, 9(6), 748-761; Montalvo, W., and E. Larson. (2014). Participant comprehension of research for which they volunteer: A systematic review. *Journal of Nursing Scholarship*, 46(6), 423-431.

[287] Jansen, L.A., et al. (2011). Unrealistic optimism in early phase oncology trials. *IRB*, 33(1), 1-8.

[288] Presentations to the Secretary's Advisory Committee for Human Research Protections (SACHRP). (2014, July 21). Retrieved February 2, 2015 from http://videocast.nih.gov/summary.asp?Live=14429; Flory, J., and E. Emanuel. (2004). Interventions to improve research participants' understanding in informed consent for research: A systematic review. *JAMA*, 292(13), 1593-1601.

[289] Nishimura, A., et al. (2013). Improving understanding in the research informed consent process: A systematic review of 54 interventions tested in randomized control trials. *BMC Medical Ethics*, 14(28), 1-15.

[290] Galbraith, K.L., Keck, A.S., and C. Little. (2014). Single-site community consultation for emergency research in a community hospital setting. *Prehospital Emergency Care*, 18(3), 328-334; Callaway, C.W. (2014). Studying community consultation in exception from informed consent trials. *Critical Care Medicine*, 42(2), 451-453.

[291] Labuzetta, J.N., Burnstein, R., and J. Pickard, *supra* note 8; Grisso, T., and L. Steinberg, *supra* note 8.

[292] Mastroianni, A., and J. Kahn, *supra* note 197.

[293] SACHRP, *supra* note 220.

[294] Anderson, J.A., Eikholdt, M. and J. Illes. (2013). Neuroethical issues in clinical neuroscience research. In J.L. Bernat and R. Bereford (Eds.). *Handbook of Clinical Neurology: Ethical and Legal Issues in Neurology*, Vol. 118 (pp. 336-343). Amsterdam, The Netherlands: Elsevier B.V.

[295] *Protection of Human Subjects, HHS*. 45 C.F.R. § 46.102, 46.111.

[296] Office for Human Research Protections (OHRP), HHS. (2011). Who Can Be a Legally Authorized Representative (LAR) for the Purpose of Providing Consent on Behalf of a Prospective Subject? [Webpage]. Retrieved February 2, 2015 from http://www.hhs.gov/ohrp/policy/faq/informed-consent/legally-authorized-representative-for-providing-consent.html.

[297] OHRP has investigated institutions that have allowed research to proceed using LARs that were not explicitly authorized under state law. See Letter from Kristina C. Borror, Compliance Oversight Coordinator, Division of Compliance Oversight, Office for Human Research Protections (OHRP), to Nathan Kase, Dean (Interim), Mount Sinai School of Medicine. (2002, May 7). RE: Human Research Subject Protections Under Multiple

Project Assurance (MPA) M-1155. Retrieved February 2, 2015 from http://www.hhs.gov/ohrp/detrm_letrs/YR02/may02a.pdf.
[298] SACHRP, *supra* note 220.
[299] See generally, Tovino, S.A., *supra* note 211.
[300] Bailey, R.W. (2003). *Rogue Scholar: The Sinister Life and Celebrated Death of Edward H. Rulloff*. Ann Arbor, MI: The University of Michigan Press.
[301] PBS, *supra* note 25.
[302] Meixner, J.B., Jr. (2015). Applications of neuroscience in criminal law: Legal and methodological issues. *Current Neurology and Neuroscience Reports*, 15(2), 513; Figdor, C. (2013). What is the "cognitive" in cognitive neuroscience? *Neuroethics*, 6(1), 105-114.
[303] Elbert, J.M. (2012). A mindful military: Linking brain and behavior through neuroscience at court-martial. *The Army Lawyer*, September, 4-24.
[304] Greely, H.T. (2009). Keynote address: Law and the revolution in neuroscience: An early look at the field. *Akron Law Review*, 42, 687-713; *People v. Varnum*, 2011 Cal. LEXIS 1558.
[305] *Roper v. Simmons*, 543 U.S. 551 (2005); *Graham v. Florida*, 560 U.S. 46(2010).
[306] Casey, P.M., et al. (2012). *Helping Courts Address Implicit Bias: Resources for Education*. National Center for State Courts. Retrieved February 2, 2015 from http://www.ncsc.org/IBReport; Werner, N.S., Kühnel, S., and H.J. Markowitsch. (2013). The neuroscience of face processing and identification in eyewitnesses and offenders. *Frontiers in Behavioral Neuroscience*, 7, article 189.
[307] Church, D.J. (2012). Neuroscience in the courtroom: An international concern. *William & Mary Law Review*, 53(5), 1824-1854.
[308] Rosen, J. (2007, March 11). The brain on the stand: How neuroscience is transforming the legal system. *New York Times*. Retrieved February 2, 2015 from http://www.nytimes.com/2007/03/11/magazine/11Neurolaw.t.html?pagewanted=all&_r=0.
[309] Farah, M.J., et al. (2009). Brain imaging and brain privacy: A realistic concern? *Journal of Cognitive Neuroscience*, 21(1), 119-127.
[310] Ibid.
[311] Jones, O.D., et al., *supra* note 11.
[312] Appelbaum, P.S. (2005). Behavioral genetics and the punishment of crime. *Psychiatric Services*, 56(1), 25-27.
[313] Church, D.J., *supra* note 307.
[314] President's Council on Bioethics. (2004). Staff Working Paper: An Overview of the Impact of Neuroscience Evidence in Criminal Law. Retrieved February 2, 2015 from http://bioethics.georgetown.edu/pcbe/background/neuroscience_evidence.html.
[315] Aspinwall, L.G., Brown, T.R., and J. Tabery. (2012). The double-edged sword: Does biomechanism increase or decrease Judges' sentencing of psychopaths? *Science*, 337(6096), 846-849; Carey, B. (2012, August 16). Study of judges finds evidence from brain scans led to lighter sentences. *New York Times*. Retrieved February 2, 2015 from http://www.nytimes.com/2012/08/17/science/brain-evidence-sways-sentencing-in-study-of-judges.html?pagewanted=all.
[316] Munro, G.D., and C.A. Munro. (2014). "Soft" versus "hard" psychological science: Biased evaluations of scientific evidence that threatens or supports a strongly held political identity. *Basic and Applied Social Psychology*, 36(6), 533-543.
[317] Farahany, N., Database 2014. On file at Duke University. A total of 1800 judicial opinions (majority, plurality, concurrence, dissent) issued during 2005–2012 were included. Graph and analysis based on 1586 majority and plurality opinions only.
[318] Kendler, K.S., and L. Eaves. (2005). *Psychiatric Genetics*. Washington, DC: American Psychiatric Publishing.
[319] *Roper v. Simmons*, 543 U.S. 551 (2005); *Graham v. Florida*, 560 U.S. 46 (2010).
[320] Farahany, N., *supra* note 317.
[321] U.S. Constitution, amendment VIII; U.S. Constitution, amendment XIV; *Pate v. Robinson*, 383 U.S. 375, 376 (1966); *Dusky v. United States*, 362 U.S. 402 (1960) (per curiam). Competence to stand trial and be sentenced is a legal standard that should not be conflated with lay use of the terms "competency" or "capacity."
[322] *Offenders with Mental Disease or Defect*. 18 U.S.C. § 4241.
[323] Aronson, J.D. (2007). Brain imaging, culpability, and the juvenile death penalty. *Psychology, Public Policy, and Law*, 13(2), 115-142; Shen, F.X. (2013). Neuroscience, mental privacy, and the law. *Harvard Journal of Law and Public Policy*, 36, 653-713.
[324] Elbert, J.M., *supra* note 303.
[325] *United States v. Rothman*, No. 08-20895-CR-UNGARO/SIMONTON, 2010 U.S. Dist. LEXIS 127639 (S.D. Fla. Aug. 18, 2010).
[326] Greely, H.T. (2013). Mind Reading, Neuroscience, and the Law. In S.J. Morse and A.L. Roskies. (Eds.). *A Primer on Criminal Law and Neuroscience: A Contribution of the Law and Neuroscience Project, Supported by the MacArthur Foundation* (pp. 120-149). New York, NY: Oxford University Press.
[327] Dressler, J. (2009). *Understanding Criminal Law*. LexisNexis: Matthew Bender, p. 117.
[328] Greely, H.T., *supra* note 304.

[329] Greene, J., and J. Cohen. (2004). For the law, neuroscience changes nothing and everything. *Philosophical Transactions of the Royal Society of London, Series B: Biological Sciences*, 359(1451), 1775-1785; Morse, S.J. (2008). Determinism and the death of folk psychology: Two challenges to responsibility from neuroscience. *Minnesota Journal of Law, Science & Technology*, 9(1), 1-36; Roskies, A.R. (2006). Neuroscientific challenges to free will and responsibility. *Trends in Cognitive Sciences*, 10(9), 419-423.

[330] Baumeister, R.F., Masicampo, E.J., and C.N. DeWall. (2009). Prosocial benefits of feeling free: Disbelief in free will increases aggression and reduces helpfulness. *Personality and Social Psychology Bulletin*, 35(2), 260-268; Vohs, K.D., and J.W. Schooler. (2008). The value of believing in free will: Encouraging a belief in determinism increases cheating. *Psychological Science*, 19(1), 49-54.

[331] Model Penal Code § 2.01. Requirement of Voluntary Act; Omission as Basis of Liability; Possession as an Act. (2014).

[332] Libet, B. (1999). Do we have free will? *Journal of Consciousness Studies*, 6(8-9), 47-57; Greene, J., and J. Cohen, *supra* note 329; Schleim, S. (2012). Brains in context in the neurolaw debate: The examples of free will and "dangerous" brains. *International Journal of Law and Psychiatry*, 35(2), 104-111.

[333] *Varnum v. Brazelton*, 2012 U.S. Dist. LEXIS 96979 (C.D. Cal. June 15, 2012).

[334] Aronson, J.D. (2010). The law's use of brain evidence. *Annual Review of Law and Social Science*, 6, 93-108; Snead, O.C. (2007). Neuroimaging and the "complexity" of capital punishment. *NYU Law Review*, 82, 1265-1339.

[335] Rushing, S.E. (2012). The admissibility of brain scans in criminal trials: The case of positron emission tomography. *Court Review*, 50(2), 62-69; Edersheim, J.G., Brendel, R.W., and B.H. Price. (2012). Neuroimaging, Diminished Capacity and Mitigation. In J.R. Simpson. (Ed.). *Neuroimaging in Forensic Psychiatry: From the Clinic to the Courtroom*, First Edition (pp. 163-193). Chichester, UK: Wiley-Blackwell.

[336] *Roper v. Simmons*, 543 U.S. 551 (2005); *Graham v. Florida*, 560 U.S. 46 (2010).

[337] *Roper v. Simmons*, 543 U.S. 551 (2005).

[338] *Graham v. Florida*, 560 U.S. 46 (2010); *Miller v. Alabama*, 132 U.S. 2455 (2012).

[339] See, e.g., Morse, S.J. (2006). Brain overclaim syndrome and criminal responsibility: A diagnostic note. *Ohio State Journal of Criminal Law*, 3, 397-412.

[340] Farahany, N.A., and J.E. Coleman, Jr. (2009). Genetics, Neuroscience, and Responsibility. In N.A. Farahany (Ed.). *The Impact of Behavioral Science on Criminal Law* (pp. 183-240). New York, NY: Oxford University Press, pp. 202-205.

[341] Jones, O.D., and F.X. Shen. (2012). Law and neuroscience in the United States. In T.M. Spranger, (Ed.). *International Neurolaw: A Comparative Analysis*, (p. 349). New York, NY: Springer-Verlag.

[342] Aharoni, E., et al. (2013). Neuroprediction of future rearrest. *Proceedings of the National Academy of Sciences*, 110(15), 6223-6228. Rearrest is merely proxy for recidivism because an individual can commit a crime without being arrested.

[343] Kiehl, K.A., Liddle P.F., and J.B. Hopfinger. (2000). Error processing and the rostral anterior cingulate: An event related fMRI study. *Psychophysiology*, 37(2), 216–223.

[344] Greely, H.T., *supra* note 304.

[345] Kiehl, K.A. (2008). Without Morals: The Cognitive Neuroscience of Criminal Psychopaths. In W. Sinnott-Armstrong (Ed.). *Moral Psychology, Volume 3: The Neuroscience of Morality: Emotion, Brain Disorders, and Development* (pp. 119-149). Cambridge, MA: MIT Press.

[346] Eastman, N., and C. Campbell. (2006). Neuroscience and legal determination of criminal responsibility. *Nature Reviews Neuroscience*, 7(4), 311-318.

[347] Aharoni, E., et al., *supra* note 342; Jones, O.D., et al. (2013). Law and neuroscience. *The Journal of Neuroscience*, 33(45), 17624-17630.

[348] Greely, H.T. (2008). Neuroscience and criminal justice: Not responsibility but treatment. *University of Kansas Law Review*, 56, 1103-1138.

[349] Depo-Provera® is produced by Pfizer, New York, New York.

[350] Greely, H.T. (2012). Direct brain interventions to "treat" disfavored human behaviors: Ethical and social issues. *Clinical Pharmacology & Therapeutics*, 91(2), 163-165.

[351] See, e.g., *McNeel v. Union Pacific R. Co.*, 753 N.W.2d 321 (Neb. 2008).) (Plaintiff brought an action under the Federal Employer's Liability Act alleging injuries from inhalation of fumes while employed as a railroad conductor. Plaintiff sought to introduce single photon emission computed tomography (SPECT) scans showing that he suffered from toxic encephalopathy from an unknown toxin); Murphy, E.R., and H.T. Greely. (2011). What will be the limits of neuroscience-based mindreading in the law? In J. Illes and B.J. Sahakian. (Eds.) *The Oxford Handbook of Neuroethics* (pp. 635-654). New York, NY: Oxford University Press.

[352] Wagner, T.D., et al. (2013). An fMRI-based neurologic signature of physical pain. *New England Journal of Medicine*, 368(15), 1388-1397.

[353] Murphy, E.R., and H.T. Greely, *supra* note 351.

[354] *Van Middlesworth v. Century Bank*, No. 215512, 2000 Mich. App. LEXIS 2369 (Mich Ct. App., May 5, 2000).

[355] See, e.g., In re H.P., Np. 07JC19926, 2009-Ohio-2186. (A single mother's parental rights were terminated, in part, due to her traumatic brain injury and resulting cognitive impairments); *In re G.V.*, 674 S.E.2d 479(N.C.

Ct. App. 2009) (Neurological evidence about a child's development was used to determine whether termination of parental rights was in the best interest of the child).

[356] Mitchell, T.L., et al. (2005). Racial bias in mock juror decision-making: A meta-analytic review of defendant treatment. *Law and Human Behavior*, 29(6), 621-637.

[357] Danziger, S., Levav, J., and L. Avnaim-Pesso. (2011). Extraneous factors in judicial decisions. *Proceedings of the National Academy of Sciences*, 108(17), 6889-6892.

[358] Casey, P.M., et al., *supra* note 306.

[359] Werner, N.S., Kühnel, S., and H.J. Markowitsch, *supra* note 306.

[360] Cohen, A.O., and B.J. Casey. (2014). Rewiring juvenile justice: The intersection of developmental neuroscience and legal policy. *Trends in Cognitive Sciences*, 18(2), 63-65.

[361] Rauscher, F.H., Shaw, G.L., and K.N. Ky. (1993). Music and spatial task performance. *Nature*, 365(6447), 611.

[362] Bangerter, A., and C. Heath. (2004). The Mozart effect: Tracking the evolution of a scientific legend. *British Journal of Social Psychology*, 43(4), 605-623; Chabris, C.F. (1999). Prelude or requiem for the 'Mozart effect'? *Nature*, 400(6747), 826-827; Pietschnig, J., Voracek, M., and A.K. Formann. (2010). Mozart effect-Schmozart effect: A meta-analysis. *Intelligence*, 38(3), 314-323; Steele, K.M., Brown, J.D., and J.A. Stoecker. (1999). Failure to confirm the Rauscher and Shaw description of recovery of the Mozart effect. *Perceptual and Motor Skills*, 88(3), 843-848.

[363] DiPietro, J.A. (2000). Baby and the brain: Advances in child development. *Annual Review of Public Health*, 21, 455-471.

[364] Rosen, J., *supra* note 308.

[365] Farah, M.J., et al., *supra* note 309.

[366] Farahany, N.A. (2011). Incriminating thoughts. *Stanford Law Review*, 64, 351-408; Bizzi, E., et al. (2009). Using imaging to identify deceit: Scientific and ethical questions. *American Academy of Arts and Sciences*, Retrieved February 3, 2015 from http://www.amacad.org/pdfs/deceit.pdf; Meixner, J.B. (2012). Liar liar, jury's the trier? The future of neuroscience-based credibility assessment in the court. *Northwestern University Law Review*, 106(3), 1451-1488; Miyawaki, Y., et al. (2008). Visual image reconstruction from human brain activity using a combination of multiscale local image decoders. *Neuron*, 60(5), 915–929.

[367] Farah, M.J., et al., *supra* note 309.

[368] Formisano, E., et al. (2008). "Who" is saying "what"? Brain-based decoding of human voice and speech. *Science*, 322(5903), 970-973; Andics, A., et al. (2010). Neural mechanisms for voice recognition. *Neuroimage*, 52(4), 1528-1540; Bizzi, E., et al., *supra* note 366; Farahany, N.A., *supra* note 366.

[369] Farahany, N.A., *supra* note 366; Rissman, J., Greely, H.T., and A.D. Wagner. (2010). Detecting individual memories through the neural decoding of memory states and past experience. *Proceedings of the National Academy of Sciences*, 107(21), 9849-9854; Farwell, L.A., Richardson, D.C., and G.M. Richardson. (2013). Brain fingerprinting field studies comparing P300-MERMER and P300 brainwave responses in the detection of concealed information. *Cognitive Neurodynamics*, 7(4), 263-299.

[370] See Nishimoto, S., et al. (2011). Reconstructing visual experiences from brain activity evoked by natural movies. *Current Biology*, 21(19), 1641-1646.

[371] Farahany, N.A., *supra* note 366; Simpson, J.R. (2008). Functional MRI lie detection: Too good to be true? *Journal of the American Academy of Psychiatry and the Law Online*, 36(4), 491-498; Bizzi, E., et al., *supra* note 366.

[372] Farahany, N.A. (2008, April 13). The government is trying to wrap its mind around yours. *The Washington Post*. Retrieved February 3, 2015 from http://www.washingtonpost.com/wp-dyn/content/article/2008/04/11/AR2008041103296.html; Farah, M.J., et al. (2014). Functional MRI-based lie detection: Scientific and societal challenges. *Nature Reviews Neuroscience*, 15(2), 123-131.

[373] Giridharadas, A. (2008, September 8). Brain scan a new wave in criminal evidence. *The International Herald Tribune*.

[374] See Farahany, N.A. (2012). Searching Secrets. *University of Pennsylvania Law Review*, 160, 1239-1308; Brennan-Marquez, K. (2013). A modest defense of mind reading. *Yale Journal of Law and Technology*, 15(2), 214-272; Shen, F.X., *supra* note 323.

[375] See Farahany, N.A., *supra* note 366; Brennan-Marquez, K., *supra* note 374; Shen, F.X., *supra* note 323.

[376] Greely, H.T., *supra* note 304.

[377] Greene, J., and J. Cohen, *supra* note 332; Greene, J.D., John and Ruth Hazel Associate Professor of Psychology; Director, Moral Cognition Lab, Harvard University. (2014). Potential Implications of Advances in Neuroscience Research for Ethics and Moral Decision Making. Presentation to the Presidential Commission for the Study of Bioethical Issues, June 10. Retrieved February 4, 2015 from http://bioethics.gov/node/3778.

[378] Menninger, K. (1968). *The Crime of Punishment*. New York, NY: The Viking Press.

[379] Morse, S.J., and W.T. Newsome. (2013). Criminal Responsibility Criminal Competence, and Prediction of Criminal Behavior. In S.J. Morse and A.L. Roskies. (Eds.). *A Primer on Criminal Law and Neuroscience: Contribution of the Law and Neuroscience Project, Supported by the MacArthur Foundation* (pp. 150-178). New York, NY: Oxford University Press.

[380] Farahany, N.A. (2012). A neurological foundation for freedom. *Stanford Technology Law Review*, 4, 1-15.

[381] Morse, S.J., *supra* note 329.
[382] Morse, S.J., and W.T. Newsome, *supra* note 379.
[383] Morse, S.J., and W.T. Newsome, *supra* note 379, p. 159.
[384] Ibid, p. 160.
[385] Ibid, p. 162.
[386] Ibid, pp. 174-175.
[387] Grassian, S. (2006). Psychiatric effects of solitary confinement. *Journal of Law & Policy*, 22, 325-383; Smith, P.S. (2006). The effects of solitary confinement on prison inmates: A brief history and review of the literature. *Crime and Justice*, 34(1), 441-528.
[388] Lovell, D. (2008). Patterns of disturbed behavior in a supermax population. *Criminal Justice and Behavior*, 35(8), 985-1004; Pizarro, J., and V.M.K. Stenius. (2004). Supermax prisons: Their rise, current practices, and effect on inmates. *The Prison Journal*, 84(2), 248-264.
[389] Fazel, S., and K. Seewald. (2012). Severe mental illness in 33588 prisoners worldwide: Systematic review and meta-regression analysis. *The British Journal of Psychiatry*, 200(5), 364-373; Fazel, S., and J. Danesh. (2002). Serious mental disorder in 23,000 prisoners: A systematic review of 62 surveys. *The Lancet*, 359(9300), 545-550; Shiroma, E.J., Ferguson, P.L., and E.E. Pikelsimer. (2012). Prevalence of traumatic brain injury in an offender population: A meta-analysis. *Journal of Head Trauma Rehabilitation*, 27(3), E1-E10.
[390] Kaba, F., et al. (2014). Solitary confinement and risk of self-harm among jail inmates. *American Journal of Public Health*, 104(3), 442-447.
[391] Fazel, S., and K. Seewald, *supra* note 389.
[392] Ravindran, S. (2014, February 27). Twilight in the box: The suicide statistics, squalor & recidivism haven't ended solitary confinement. Maybe the brain studies will. *Aeon*. Retrieved February 4, 2015 from http://aeon.co/magazine/society/what-solitary-confinement-does-to-the-brain/; Willigan, M. (2014, August 4). What solitary confinement does to the human brain. *Solitary Watch*. Retrieved February 4, 2015 from http://solitarywatch.com/2014/08/04/what-solitary-confinement-does-to-the-human-brain/; Ramlagan, N. (2014, February 15). Solitary confinement fundamentally alters the brain, scientists say. *American Association for the Advancement of Science*. Retrieved February 4, 2015 from http://www.aaas.org/news/solitary-confinement-fundamentally-alters-brain-scientists-say.
[393] Presidential Commission for the Study of Bioethical Issues. (2014). Law and Neuroscience Framing and Possible Recommendations. Member Discussion, November 5. Retrieved February 4, 2015 from http://bioethics.gov/node/4319.
[394] Morse, S.J., and W.T. Newsome, *supra* note 379.
[395] Morse, S.J. (2011). Avoiding irrational neurolaw exuberance: A plea for neuromodesty. *Mercer Law Review*, 62, 837-859.
[396] Ibid, p. 850.
[397] Ibid.
[398] The Royal Society (2011). *Brain Waves Module 4: Neuroscience and the Law* (7). London: The Royal Society. Retrieved February 4, 2015 from https://royalsociety.org/policy/projects/brain-waves/responsibility-law/.
[399] Faigman, D.L., Monahan, J., and C. Slobogin. (2014). Group to individual (G2i) inference in scientific expert testimony. *The University of Chicago Law Review*, 81(2), 417-480.
[400] Aue, T., Lavelle, L.A., and J.T. Cacioppo. (2009). Great expectations: What can fMRI research tell us about psychological phenomena? *International Journal of Psychophysiology*, 73(1), 10-16.
[401] Monteleone, G.T., et al. (2009). Detection of deception using fMRI: Better than chance, but well below perfection. *Social Neuroscience*, 4(6), 528-538.
[402] A notable exception is in Supreme Court cases where the Court has carved out categorical exceptions to the applicability of the death penalty for mentally retarded offenders and juveniles. See Farahany, N.A. (2009). Cruel and Unequal Punishments. *Washington University Law Review*, 86, 859-915.
[403] Faigman, D.L., Monahan, J., and C. Slobogin, *supra* note 399.
[404] *Ford v. Wainwright*, 477 U.S. 399 (1986).
[405] *Kemp v. Cockrell*, No. 4:99-CV-490-Y, 2003 WL 21448584 (N.D. Tex. 2003); Ayres, P. (2014). Medicate to execute: The dilemma of involuntary medication and mentally ill death row inmates. *The Forum*, 1(1), 28-46.
[406] Greely, H.T., *supra* note 350.
[407] Greely, H.T., *supra* note 348.
[408] McCabe, D.P., and A.D. Castel. (2008). Seeing is believing: The effect of brain images on judgments of scientific reasoning. *Cognition*, 107(1), 343-352; Weisberg, D.S., et al. (2008). The seductive allure of neuroscience explanations. *Journal of Cognitive Neuroscience*, 20(3), 470-477; Rhodes, R.E., Rodriguez, F., and P. Shah. (2014). Explaining the alluring influence of neuroscience information on scientific reasoning. *Journal of Experimental Psychology: Learning, Memory, and Cognition*, 40(5), 1432-1440; McCabe, D.P., Castel, A.D., and M.G. Rhodes. (2011). The influence of fMRI lie detection evidence on juror decision-making. *Behavioral Sciences and the Law*, 29(4), 566-577; Schweitzer, N.J., and M.J. Saks. (2011). Neuroimage evidence and the insanity defense. *Behavioral Sciences and the Law*, 29(4), 592-607.

[409] Brainfacts.org is sponsored by the Kavli Foundation, the Gatsby Foundation, and the Society for Neuroscience. The Kavli Foundation, The Gatsby Foundation, and the Society for Neuroscience, *supra* note 142.

[410] Weisberg, D.S., et al., *supra* note 408.

[411] For these seminars, AAAS partners with the Dana Foundation, the Judicial Division of the American Bar Association, the National Center for State Courts, and the Federal Judicial Center. American Association for the Advancement of Science (AAAS). (2014). Judicial seminars on emerging issues in neuroscience [Webpage]. Retrieved February 4, 2015 from http://www.aaas.org/page/judicial-seminars-emerging-issues-neuroscience.

[412] The MacArthur Foundation Research Network on Law and Neuroscience. (2014, March 31). Comments submitted to the Presidential Commission for the Study of Bioethical Issues, p. 3.

[413] MacArthur Foundation Research Network on Law and Neuroscience. (n.d.). Education and outreach [Webpage]. Retrieved February 4, 2015 from http://www.lawneuro.org/outreach.php; Integrative Law Institute. (n.d.). Law practice and the human brain: Neuro-literacy 101 for lawyers, mediators, and judges [Webpage]. Retrieved February 4, 2015 from http://events.r20.constantcontact.com/register/event?oeidk=a07e6hgkejb 2f97cb18&llr=6rtybkkab; American Academy of Psychiatry and the Law. (n.d.). Forensic psychiatry review course [Webpage]. Retrieved February 4, 2015 from http://www.aapl.org/review.htm; Casey, P.M., et al., *supra* note 306.

[414] National Research Council Committee on DNA Forensic Science. (1992). *DNA Technology in Forensic Science*. Washington, DC: The National Academies Press. Retrieved February 4, 2015 from http://www.nap.edu/catalog/1866/dna-technology-in-forensic-science; National Research Council Committee on DNA Forensic Science. (1996). *The Evaluation of Forensic DNA Evidence: An Update*. Washington, DC: The National Academies Press. Retrieved February 4, 2015 from http://www.nap.edu/catalog/5141/the-evaluation-of-forensic-dna-evidence.

[415] National Research Council Committee to Review the Scientific Evidence on the Polygraph. (2003). *The Polygraph and Lie Detection*. Washington, DC: The National Academies Press.

[416] Social Security Administration. (1996). SSR 96-7p: Policy Interpretation Ruling: Titles II and XVI: Evaluation of Symptoms in Disability Claims: Assessing the Credibility of an Individual's Statements. Retrieved February 4, 2015 from http://www.socialsecurity.gov/OP_Home/rulings/di/01/SSR96-07-di-01.html.

[417] Rusconi, E., and T. Mitchener-Nissen, *supra* note 10.

[418] Greely, H.T., *supra* note 304.

[419] Ibid.

[420] Jones, O.D., et al., *supra* note 11.

[421] Ibid.

[422] Ibid.

[423] Ibid.

[424] HIV Modelling Consortium. (2011). About us [Webpage]. Retrieved February 6, 2015 from http://www.hivmodelling.org/about-us.

[425] PCSBI, *supra* note 1.

INDEX

C

D

E

F

G

H

I

N

O

P

Q

R

S

T

U

V

W

Y